Bakhytzhan Kedelbaev
Kulzada Lakhanova
Darikha Kudasova

Hydrolytic hydrolysis and hydrogenation of carbon-containing raw materials

AF625227

ScienciaScripts

Imprint
Any brand names and product names mentioned in this book are subject to trademark, brand or patent protection and are trademarks or registered trademarks of their respective holders. The use of brand names, product names, common names, trade names, product descriptions etc. even without a particular marking in this work is in no way to be construed to mean that such names may be regarded as unrestricted in respect of trademark and brand protection legislation and could thus be used by anyone.

Cover image: www.ingimage.com

This book is a translation from the original published under ISBN 978-620-2-06822-2.

Publisher:
Sciencia Scripts
is a trademark of
Dodo Books Indian Ocean Ltd. and OmniScriptum S.R.L publishing group

120 High Road, East Finchley, London, N2 9ED, United Kingdom
Str. Armeneasca 28/1, office 1, Chisinau MD-2012, Republic of Moldova, Europe
Printed at: see last page
ISBN: 978-620-7-95530-5

Copyright © Bakhytzhan Kedelbaev, Kulzada Lakhanova, Darikha Kudasova
Copyright © 2024 Dodo Books Indian Ocean Ltd. and OmniScriptum S.R.L publishing group

Bakhytzhan Kedelbaev
Kulzada Lakhanova
Darikha Kudasova

Hydrolytic hydrolysis and hydrogenation of carbon-containing raw materials

CONTENTS.

REFERENCE.

Report 100c., 23 tables, 25 figures, sources 106,

cellulose, guza pai, beer pellets, hydrolytic hydrolysis and hydrogenation, glucose, xylose, sorbitol, xylitol, combined (hybrid) process.

The object of the study is carbohydrate-containing plant raw materials and wastes, including inm2016 beer pellets and guza-pai pulp.

The aim of the work is technical substantiation and development of technology for deep processing of carbohydrate-containing raw materials to produce xylitol and sorbitol by means of chemical and enzymatic hydrolytic hydrolysis and hydrogenation of carbohydrate-containing plant raw materials and wastes, based on the use of a combined (hybrid) process.

The methodology of work is based on the development of technology of chemical and enzymatic hydrolytic hydrolysis and hydrogenation of carbohydrate-containing plant raw materials and wastes (beer pellets and cellulose guza-paya), based on the use of a combined (hybrid) method.

The research on studying the process of chemical hydrolytic hydrolysis and hydrogenation of cellulose of guza-paya and beer pellets was carried out, the optimal parameters of the process were developed. The process of enzymatic hydrolytic hydrolysis and hydrogenation of beer pellet was investigated. As a result, we realized a combined (hybrid) hydrolysis-hydrogenation process of sorbitol and xylitol production. New generation catalysts for these processes were developed and their activity was investigated.

The influence of experiment temperature, pH, hydrogen pressure, nature of the acid used, reducing agent, amount and 3 nature of the applied catalyst, ferroalloy promoter on polysaccharide conversion and selectivity for sorbitol, xylitol and mannitol.

The developed processes allow to improve traditional processes in terms of elimination of numerous stages of purification and isolation of intermediate products. They make it possible to realize a single-reactor combined (hybrid) process for the production of such valuable chemicals as xylitol and sorbitol.

The research results will be used in the next stage of work to determine the physicochemical properties of the contacts and to identify the correlation between catalytic, adsorption and physicochemical properties; as well as to study the kinetics and mechanism of processes, quantum-chemical calculations and modeling, which will allow us to develop original technologies for the production of xylitol and sorbitol, taking into account the requirements of the technical task.

According to the results of the second stage of research, there is a business partner of SF LLP NPF "Anazher" to implement the technology of xylitol and sorbitol production by combined (hybrid) chemical and enzymatic hydrolytic hydrolysis and hydrogenation of cellulose of guza-paya and beer pellet.

INTRODUCTION

Carbohydrate-containing plant raw materials and wastes are very promising alternative resources for the production of valuable chemical compounds. Their advantages lie in their renewability, or even practical inexhaustibility [1-4]. Of particular interest is the search for catalytic technologies of a single-stage, combined (hybrid) process of obtaining valuable substances directly from these raw materials, excluding the technological stages of separation and purification of intermediates. One-stage process organization makes it possible to obtain such compounds as xylitol and sorbitol from plant polysaccharide by hydrolysis-hydrogenation reaction. Intensive research in this direction is carried out in the world. A wide range of catalytic systems for hydrolytic transformations of renewable polysaccharide, such as catalytic systems based on carbon, oxides, zeolites, ion-exchange resins have been proposed. Various technological schemes for the realization of such processes have been developed, and different variations of methods for one-stage processing of polysaccharides of carbohydrate-containing plant raw materials and wastes have been proposed. However, the literature data published to date on the use of various catalytic systems in the transformation of these polysaccharides often contradict each other, and the proposed catalysts show low activity or stability. It should be noted that most researchers have studied exclusively the hydrolysis of polysaccharides into glucose, and the number of works devoted to the possibility of obtaining, for example, from beer pellets and cellulose guza-pai xylitol and sorbitol as a result of combined (hybrid) processes is extremely small. Although, these compounds are substances extremely demanded in many modern branches of chemical, food, perfume, medical and other industries. In particular, sorbitol and xylitol are widely used in the production of surfactants, synthetic resins, varnishes, olive oil, vitamin C, food products, etc.. A significant increase in the number of scientific publications devoted to one-stage processing of biomass components and especially polysaccharides over the last 10 years indicates the high relevance of the problem of its transformation into valuable chemicals such as sorbitol and xylitol. To date, a large amount of experimental data on the hydrolysis of cellulose to glucose has been obtained, but information on one-step combined (hybrid) processes for the production of sorbitol and xylitol is insufficient, and the results described in the literature are often inconsistent with each other.

Analysis of the literature on the topic of work has shown that in the studied area there are a number of unsolved problems, the search for the solution of which seems useful for the creation of effective catalytic technologies for the transformation of hydrocarbon-containing plant raw materials and wastes into valuable chemicals.

Relevance.

At present, the efforts of many scientific teams are focused on the development of new, efficient and environmentally safe ways to produce polyatomic alcohols from plant biomass. Its huge natural renewable reserves can fully satisfy the needs of modern industry in polyatomic alcohols, which is especially relevant given the prospect of gradual depletion of hydrocarbon fossil reserves. Although the one-step combined (hybrid) process of hydrolysis-hydrogenation of polysaccharides to polyatomic alcohols in the presence of heterogeneous metal-containing catalysts has been known for quite a long time, however, there are practically no studies in the literature aimed at studying its physicochemical basis. Obviously, the results of such studies can contribute to the improvement of polysaccharide conversion efficiency and become a fundamental basis for new, improved technologies of biomass processing into valuable chemicals. Thus, it should be noted that if even a small fraction of the above-mentioned sources are converted into useful products by chemical or enzymatic means, this will provide a very tangible and, importantly, renewable source of carbohydrates, substitutes for petroleum products, as well as initial compounds for microbiological synthesis. The above circumstances determine the relevance of scientific research aimed at creating new technologies for chemical and enzymatic hydrolytic hydrolysis and hydrogenation of polysaccharides of plant raw materials to polyatomic alcohols.

Thus, the processing of renewable hydrocarbon-containing plant raw materials and wastes into industrially important chemicals by a single-stage combined (hybrid) process is of great practical and theoretical interest.

Novelty of the work.

The technology of one-stage combined (hybrid) process of chemical hydrolytic hydrolysis and hydrogenation of cellulose of guza-paya and beer gum has been developed, the optimum parameters of the process have been revealed. The effect of experiment temperature, pH, hydrogen pressure, nature of acid used, reducing agent, amount and nature of applied catalyst, ferroalloy promoter additive on polysaccharide conversion and selectivity for sorbitol, xylitol and mannitol was investigated. The process of enzymatic hydrolytic hydrolysis and hydrogenation of beer pellets was systematically investigated. As a result, we have developed a process of combined (hybrid) hydrolysis-hydrogenation of sorbitol and xylitol production and studied the kinetics of polysaccharide depolymerization processes. New generation catalysts for these processes were developed.

Practical value of the work

The developed one-stage combined (hybrid) technology of chemical hydrolytic hydrolysis and hydrogenation of cellulose of guza-pai and beer pellets can be successfully used in the organization of enterprises for complex processing of hydrocarbon-containing plant raw materials and wastes. The combined (hybrid) technology of enzymatic hydrolytic hydrolysis and hydrogenation of beer pellets, for which the optimal conditions of process realization have been determined, is also very promising. The developed technologies of chemical and enzymatic hydrolytic hydrolysis and hydrogenation of guza-paya and beer pellets are ready for implementation in industrial production of related direction.

The research objectives for 2016 are:

1. Development of technology of chemical and enzymatic hydrolytic hydrolysis and hydrogenation of hydrocarbon-containing plant raw materials and wastes, based on the use of a combined (hybrid) method

2. Development of technology for synthesis of a new generation of hydrolytic and liquid-phase hydrogenation catalysts and selection of optimal process parameters in

their presence.

The results of the research will be used in the next stage of work to determine the physicochemical properties of contacts and to identify the correlation between catalytic, adsorption and physicochemical properties; as well as to study the kinetics and mechanism of processes, quantum-chemical calculations and modeling, which will allow us to develop original technologies for the production of xylitol and sorbitol, taking into account the requirements of the terms of reference. According to the results of the second stage of research, there is a business partner of the SF LLP NPF "Anazher" to implement the technology of xylitol and sorbitol production by combined (hybrid) chemical and enzymatic hydrolytic hydrolysis and hydrogenation of cellulose guza-paya and beer pellets.

1 Main part

Development of technology of chemical and enzymatic hydrolytic hydrolysis and hydrogenation of hydrocarbon-containing plant raw materials and wastes, based on the use of a combined (hybrid) method

The world's reserves of fossil organic raw materials, represented by oil, natural gas and coal, are huge, but sooner or later they will be exhausted. As an alternative to fossil fuels, renewable sources of energy and organic raw materials are increasingly being used. The most important of them is plant raw materials formed in the process of photosynthesis.

Biomass is a renewable resource and plays a role in preventing global climate warming by slowing carbon dioxide emissions. The production of chemical products from biomass considers ways to convert biomass into sorbitol, ethanol, lactic acid and other useful chemical products by enzymatic or chemical means. Currently, the main material used for bioprocessing is starch derived from corn. In terms of resources of the main structural constituents of plants that can be utilized, cellulose is present in much greater quantities than starch. However, methods of converting cellulose into chemically useful products by reducing its molecular weight (depolymerization) have not been developed, and this resource is currently virtually untapped. A large amount of research has been conducted in the area of cellulose degradation by enzymes. However, there remain important problems associated with enzymatic methods due to low reaction rates and the need to significantly increase the enzyme activity and separate it from the product. Cellulose has the advantage of being a renewable or even practically inexhaustible plant raw material. Of particular interest is the search for technologies of one-stage combined (hybrid) method of obtaining valuable substances directly from cellulose, excluding the stages of isolation and purification of products. One-stage organization makes it possible to obtain from plant polysaccharide by hydrolysis-hydrogenation of such a compound as sorbitol, which is one of the promising sources of raw materials for industry. (6). Among the possible applications of sorbitol, a sugar alcohol, three of the most important can be mentioned. The first application is as a sweetener, which is widely used in the food industry. The second area is its use as an intermediate in the synthesis of useful compounds such as isosorbide, propylene glycol, ethylene glycol, glycerol, 1,1-sorbitan and lactic acid .Isosorbide in particular is also used in modern processes such as copolymerization in the production of polyethylene terephthalate (PET)to produce polyethylenisosorbide terephthaoate (PEIT). PEIT polymer has a higher glass transition temperature than PET, so it is expected to be used for clear plastic containers that can withstand hot water. A third application is its use as an intermediate in the production of hydrogen and liquid hydrocarbons (containing mainly C_5 and C_6 -alkanes) that can be reproduced from biomass. Hydrogen is used in fuel cells and hydrocarbons are the starting material for petrochemicals.

1.1 . Chemical methods for hydrolytic hydrogenation and hydrolysis of hydrocarbon-containing plant materials and wastes

The beginning of research on the joint hydrolysis and hydrogenation processes was laid back in the USSR by Balandin A. A. et al. who subjected cellulose to hydrolytic hydrogenation in the presence of concentrated phosphoric and sulfuric acids and Ru, Pd, and Pt-containing catalysts. Combining the processes of cellulose hydrolysis and hydrogenation of the resulting glucose allows achieving high yields of sorbitol and other higher polyols. This is explained by the fact that the glucose formed during hydrolysis is hydrogenated into sorbitol at a high rate, and due to the low concentration of free glucose in the solution, the formation of by-products is reduced [5].

PalkovitsR. et al. studied the process of combined hydrolysis and hydrogenation of cellulose in the presence of dilute mineral acids [6-8]. In [8], the process was carried out under the following conditions: hydrogen pressure 50 bar, temperature 160°C, reaction time 1 or 5 hours. In this work, catalysts with 5 % content of Pt, Pd or Ru on carbon (Pt/C, Pd/C, Ru/C) in the presence of 0.5 % or 2.5 % sulfuric or phosphoric acid were used. It is important to note that regardless of the type of metal of the catalyst used, higher cellulose conversion was achieved in reactions carried out using sulfuric acid than in reactions carried out in the presence of phosphoric acid.

The use of Pt/C catalyst in combination with sulfuric acid provided the maximum cellulose conversion of 100 *%,* in the presence of Pd/C catalyst the conversion rate reached 96 %. The use of ruthenium as a catalyst provided a lower conversion value (not more than 74%). In order to investigate the effect of metal type on cellulose conversion, experiments were carried out under the same conditions but in the absence of any metal catalysts. The maximum cellulose conversion was 69% when 2.5% sulfuric acid was used and the process was carried out for 5 hours. Hence it is concluded that the presence of metal catalysts under hydrogenolysis conditions affects the depolymerization of cellulose and the breaking of C-C and C-O bonds. The effect of metal type on the process of hydrolytic hydrogenation of cellulose was also investigated by [9]. According to the studies, the activity of metals for cellulose conversion decreases in the series Pt>Pd>Ru. However, the reactions using Ru/C as a catalyst, regardless of the type of acid, are characterized by a high yield of polyols C_5 -C_6 . In the presence of Pt/Si Pd/S catalysts, low yields of polyols C5-C_6 were observed regardless of the reaction conditions. The main products in these processes are glucose and xylose.

The maximum yield of sorbitol (*33%*) was obtained when Ru/C was used as a catalyst in the presence of 2.5% sulfuric acid, and the reaction time was 1 hr.

Optimization of the balance between the two catalytic systems plays an important role for efficient and selective hydrolytic hydrogenation of cellulose. When acid catalysts dominate, the amount of free glucose increases, which causes loss of selectivity due to its degradation. When metal catalysts dominate, the loss of selectivity is caused by the hydrogenolysis of the obtained higher polyols [10].

In [11], the authors used only trace amounts of mineral acid and a zeolite-based ruthenium dual-action catalyst. Hydrolysis under the action of water and small amounts of hydrochloric acid cleaves cellulose into soluble b-glucans. Acidic properties of zeolite promote further depolymerization of b-glucans to glucose, which is rapidly hydrogenated under the action of Ru.

In the first phase of the study, the zeolites were tested in reactions with low concentrated hydrochloric acid (35 ppm, concentration constant). The pH value was 3, which is acceptable for many production processes. The highest yield of hexitols was obtained when using zeolites of the IZU group with aluminum-silicon ratio > 15.

The next step of the study was to optimize the metal to acid ratio. Reducing the metal loading from 1.8 to 0.2 wt%, the selectivity after 24-h reaction significantly increased, leading to the maximum yield of sugar alcohols equal to 63 %. Sorbitol and mannitol are the main reaction products (the proportion of sorbitol was more than 20 %). Other by-products are ethylene glycol, 1,2-propanediol, glycerol and polyol isomers C_4 -C_5 with different degrees of hydroxylation. Reducing the ruthenium loading of the pre-temperature ruthenium below 0.2 wt% reduces the selectivity towards sugar alcohols. This is due to the predominance of acidic properties over metallic properties in solution, leading to loss of glucose due to thermal and acidic degradation. This is confirmed by the fact that up to 4% of the glucose in the investigated product mixture is wasted during the intermediate reaction. Further, the glucose concentration drops again as the hydrolysis of cellulose slows down.

Increasing the concentration of hydrochloric acid (106 ppm, pH = 2.5) significantly accelerates the conversion and complete conversion occurs after 3 hours. The study also compared different catalysts. The Pt/H-ISU catalyst showed weak properties in the hydrogenation of glucose, leaving up to 25% in the intermediate stages of the process. The performance of Rn/H-IZU catalyst was compared with the traditional Yann carbon catalyst. The amount of catalyst was set so that the total amount of Ru was distributed equally in both reactions. The Ru/C catalyst provided less depolymerization of oligomers to glucose prior to hydrogenation. This emphasizes the advantage of the bifunctional Rn/H-ICU catalyst.

In [12], the technology of acid hydrolysis of beer pellets to obtain pentose hydrolysates enriched with xylose was developed. The optimal technological parameters of acid hydrolysis were found: 4.5% concentration of H_2SO_4 , temperature 100°C, treatment duration 5 hours.

In this catalytic system, hydrolysis is promoted by HCl and, to a lesser extent, the surface acidic residues of H-USY. This creates a complex of soluble intein oligomers and glucose, which are adsorbed into the zeolite pores where the acidic zeolite residues depolymerize the oligomers to glucose. The adsorbed glucose is in close proximity to the ruthenium groups in the zeolite. It can be completely eliminated by even smaller amounts of ruthenium. This explains the high selectivity in hydrogenation of excess Ru/H-USY in hydrogenolysis, since both reactions taking place on the ruthenium surface are competitive.

Catalytic systems using mineral acids show high activity and selectivity, leading to rapid conversion of cellulose to polyols, but the application of these systems is limited by the corrosive properties of the acids and the need to neutralize the resulting waste.

In [13], a new type of Ru-containing catalysts based on super cross-linked polystyrene (CHC) was proposed for the process of hydrolytic hydrogenation of cellulose in subcritical water.

The results presented in this work showed that the use of 1.0 % Ru/CHC MN 270 catalyst allows to achieve a total yield of sorbitol and mannitol on average about 50 %.Hydrolytic hydrogenation of cellulose in subcritical water using Ru-containing CHC-based catalysts can become the basis for technological processes of cellulosic biomass conversion into feedstock for chemical synthesis and industrial production of second-generation biofuels.

An alternative method of cellulose hydrolysis is the use of commercial is heteropolyacids (HPAs) [14, 15].

Heteropolyacids are complex acids whose anion is formed by two different acid-forming oxides. There are several molecules of the other one per molecule of one of them [16]. According to authors [17, 18] heteropoly acids show high selectivity for glucose in the hydrolysis of cellulose and cellobiose. The use of heteropolyacids in combination with ruthenium catalyst on carbon (Ru/C) for the conversion of cellulose to sugar alcohols was investigated by [14]. Experiments were carried out with various acids, but maximum selectivity to C_4 - C_6 sugar alcohols was shown by phosphorus-tungstate H_3 [P(W_3 Oio)$_4$] (H_3 PW)H_4 [Si(W_3 Oio)$_4$] (H_4 SiW) acids.

The maximum yield of sorbitol was approximately 30% when using acid H_3 PWs with a concentration of 17.4 mmol/L in combination with Ru/C_5 % metal content.

Both acids are reduced under the reaction conditions, as indicated by the bluish tint of the fresh reaction solution, but re-oxidized at room temperature in contact with air, accompanied by discoloration [19]. Color change of the solution does not occur in the absence of hydrogen. The reuse of GPC, titration and nuclear magnetic examination of the solution before and after the reaction did not show any noticeable change. The acids H_3 PW and H_4 SiW possess higher strength and superior thermal and hydrolytic stability compared to other heteropoly acids such as H_4 [Si(Mo_3 Oio$)_4$] and H_3 [PMoioV 0_{240}]. These properties account for the high catalytic activity of heteropolyacids H_3 PW and FCSiW in the process of cellulose hydrolysis. In the case of H_4 [Si(Mo_3 Oio$)_4$] and H_3 [PMoi V 0_{024} o], a conversion of up to 45% is achieved, but the yield of major products is below 5% [20].

The authors [21] investigated the hydrolytic hydrogenation of cellulose to sugar alcohol in the presence of acid and 5% Ru/C in water, A high yield of sugar alcohols was achieved 12 times faster. This effect is attributed to the enhanced contact between cellulose and the catalyst, which promotes a certain level of cellulose depolymerization to obtain a high yield of sugar alcohols.

The authors [22] realized the process of direct conversion of cellulose to sorbitol in the presence of a multicomponent ruthenium catalyst. The positive effect was achieved at a reaction temperature of 165 °C. No deactivation of the process under these conditions was observed.

The authors [23] studied the hydrolysis and hydrogenation of cellobiose in $ZnCI_2$ ($4H_2$ 0) solvent was studied to optimize conditions for the conversion of lignocellulose (the most abundant renewable resource) to sorbitol. Water at neutral pH does not allow hydrolysis of cellobiose under the experimental conditions, but relative hydrogenation from cellobiose to 3-p-D-glucopyranosyl-D-glucitol over a catalyst in the presence of H_2 does occur.

It was shown that the direct conversion of cellobiose to sorbitol is possible in the presence of Ru/C catalyst, temperature 125 °C for 4 hours.

The authors [29] carried out the process of direct conversion of cellulose in the presence of a bifunctional catalyst.by dual complex catalysts containing both sulfonate groups and nanoparticles. A high yield of sorbitol (71.1%)was obtained in neutral aqueous solution without liquid phase acid at an intermediate reaction temperature of 165 °C. No deactivation was observed even after 5 repeated reactions.

The authors [25] showed the possibility of direct conversion of cellulose to C_6 -sugar alcohol at a pressure of 4 MPa and a temperature of 12O° Cв in the presence of a ruthenium catalyst. This process is considered as a platform for a new generation of natural renewable energy instead of synthetic gas.

Despite their excellent catalytic properties, the separation from solution and reuse of liquid acids cause certain difficulties [26]. For this reason, the so-called insoluble non-stoichiometric cesium salts of heteropolyacids, namely CsPW and CsSiW in combination with Ru/C catalyst[27], have great potential for use in cellulose conversion processes. An aqueous solution of Cs_2 CO_3 (0.1 M) was added dropwise to an aqueous solution of heteropolyacidsH PW_{32} O4oor H_4 SiWi O_{240} (0.08 M) under vigorous stirring. The resulting white suspension was left overnight. After several cycles of centrifugation and washing with water, the suspension is dried at 60°C to form a fine powder. The dried powder is calcined for 3 hours in air at 300°C or 600°C. The corresponding samples are designated by the suffix 300 or 600, e.g., $CsSiW_{300}$ and $CsSiW_{600}$. The combination of heteropolyacids with hydrogenation catalysts such as Ru/C showed a rather high conversion efficiency of cellulose to hexitols, but the yield of sorbitol was rather low.

However, these catalysts cannot be used on an industrial scale due to the difficulty of preparation and the lengthy process time.

Despite the fact that liquid mineral acids, even under mild process conditions, have high activity for many acid-catalyzed reactions, including the hydrolysis of cellulose, their use is complicated by problems associated with the difficulty of separating the acid from the solution and the subsequent costly disposal of waste. Solid acid catalysts have a number of positive properties: high activity and selectivity, long service life and ease of recovery and reuse. Therefore, the replacement of liquid acids by active and stable solid acid catalysts is of great importance in the development of chemical processes.

Significant advances in cellulose hydrolysis have been made using solid acids such as sulfated carbonaceous acids, polymer-based acids and magnetic solid acids. Carbonaceous solid acids are considered to be the most promising catalyst for cellulose hydrolysis because they provide good access of reagents to the acid centers of SO_3 H groups. High glucose yields (up to 75 %) with 80 % selectivity were achieved at 150 °C for 24 hours. However, the separation of acid and non-hydrolyzed cellulose after completion of the process needs further investigation since these catalysts and cellulose residues have similar physical and chemical properties [28]. The use of catalysts containing magnetic groups gives glucose yields of about 50%, and the problem of separation of the catalyst and unhydrolyzed cellulose is solved by applying magnetic force [29].

Naga et al. [30-33] developed a new type of carbon-based solid acid with a high density of Brønsted acid centers^O_3 Ni UNO) by pyrolytic carbonization of sugars such as glucose, sucrose, or cellulose and then sulfiting the resulting carbons. These sulfated carbons are highly active in the hydrolysis of microcrystalline cellulose to produce water-soluble saccharides even at low reaction temperature (100°C), in contrast to conventional solid strong acid catalysts such as niobium acid, H-mordenite, and Amberlyst-15 [31]. The specific surface area of the sulfated catalyst increases.

This method for the synthesis of solid acid catalysts has been widely used to find new efficient catalysts for cellulose conversion. Opbas et al. [34. 35] used sulfated activated carbon (AC). This catalyst provided 41% glucose yield and 95% selectivity after 24 h of reaction at 150°C for cellulose hydrolysis.

In [36], experimental material on the study of acid hydrolysis of polysaccharides of cellolignin residue of larch wood for the development of technology for the production of crystalline glucose was summarized.

The technology developed on the basis of this method was implemented at the pilot plant of Milgravsky plant in Riga. Using this technology it was possible to produce 200 kg of marketable glucose and 170 kg of marketable fodder yeast from 1 ton of absolutely dry wood (a.s.d.).

Nevertheless, the above-mentioned technologies have not found their industrial application due to serious difficulties associated with acid regeneration, hardware design, and environmental safety requirements of production.

A group of researchers [37] developed a method of glucose production from cellulose-containing wastes, which provides grinding of raw materials to a fraction with particle size 0.1-1.6 mm, treatment of raw materials with a solution containing 20-30% wt. of acetic acid, 2-5% wt. of hydrogen peroxide and 1.5-2.0% wt. of sulfuric acid at a temperature of 110-120°C and hydromodule 5 for 2-3 hours. Hydrolysis with concentrated sulfuric acid is carried out at room temperature at a hydromodule of 3-5 for 1-2 hours. The mass is diluted with water to a hydromodule of 80-100 and inversion is carried out at 90-100°C for 1-2 hours.

Of great interest is the work [38] related to the processing of renewable raw materials (in particular, cellulose) into raw materials for chemical synthesis and biofuels. The method of catalytic conversion of cellulose into hexitols was proposed, which includes the process of hydrolytic hydrogenation of cellulose for 3-7 minutes at a temperature of 240-250°C at a partial pressure of hydrogen 55-65 atm and with stirring of the reaction medium in the presence of a ruthenium catalyst. As a substrate of ruthenium catalyst super cross-linked polystyrene of MN 270 grade is used, and the ruthenium content in the catalyst is from 1.0 to 1.5 wt.% of the catalyst mass. Stirring of the reaction mixture is carried out by means of a propeller stirrer, the number of revolutions of which is 580-620 rpm.

The authors [39] report a method for obtaining xylose product and cellulose for chemical processing from biomass containing xylan about 10-35% per dry matter. According to the proposed method, pre-hydrolysis of the biomass containing xylan is carried out to obtain dissolved hemicellulose hydrolysate and undissolved pre-hydrolysis biomass. Thereafter, chromatographic fractionation of the dissolved hemicellulose hydrolysate, nanofiltration or crystallization of the precipitate is carried out to obtain a xylose product having a xylose content of at least 55% per dry matter content of the xylose product. This is followed by sulfate cooking of the undissolved pre-hydrolysis biomass to produce cellulose.

The use of sulfated silicon-carbon nanocomposites increased the glucose yield up to 50% with cellulose conversion of 61% under vanalogous reaction conditions [40]. In addition, [41] reported an increase in glucose yield up to 75% and cellulose conversion of 94% using sulfated mesoporous carbon CMK-3.

The invention [42], which relates to a catalyst for hydrolysis of cellulose or hydrolysis of cellulose and recovery of hydrolysis products and a method for producing sugar alcohols from cellulose, is quite interesting. A catalyst for hydrolysis of cellulose and hydrolysis of cellulose and recovery of hydrolysis products, in which a transition metal of groups 8-11 is deposited on a solid substrate, is described. A method for producing sugar alcohols is also described, comprising: hydrolysis of cellulose in the presence of a catalyst in a hydrogen-containing atmosphere at elevated pressure and recovery of a cellulose hydrolysis product. The technical effect is easy separation of the catalyst from the product, no need for pH adjustment, neutralization of acid or alkali when obtaining alcohols, possibility of reuse of the catalyst without its activation, obtaining sugar alcohols directly from cellulose using such a catalyst.

Many other solid acid catalysts such as layered niobium molybdate ($NMbMoO_6$) [43] and mesoporous oxides [44], along with conventional solid acids such as sulfated zirconium dioxide, have been used in cellulose conversion, but they have shown low catalytic activity.

Solid acids solve the issue of process environmental friendliness, but their use for the conversion of cellulose to sorbitol and mannitol is impractical due to relatively low yields [45].

Ionic liquids are a group of organic salts that exist as a liquid at relatively low temperature (<100°C). They have a number of attractive properties such as chemical and thermal stability, non-flammability and low vapor pressure. Unlike traditional volatile organic compounds, they have been widely used as environmentally friendly solvents [46]. In addition, ionic liquids can stabilize transition metal nanoparticles by maintaining their small size, large surface area and preventing nanoparticle leaching [47].

The authors [48] investigated the process of direct conversion of cellulose into sugar alcohol. The optimal ionic solvent was selected, which showed the best results in this process. This technology solves the problems of "green" and energy-saving technology

In this work [49] a new type of Ci-containing catalysts based on super cross-linked polystyrene (СПС) was proposed for the process of hydrolytic hydrogenation of cellulose in subcritical water. The results presented in this work show that the use of 1.0 % Ci/PSS WE 270 catalyst allows to achieve a total yield of sorbitol and mannitol of about 50% on average.The hydrolytic hydrogenation of cellulose in subcritical water with the use of SPS-based i-containing catalysts can become the basis for technological processes of cellulosic biomass conversion into feedstock for chemical synthesis and industrial production of second-generation biofuels.

In [50-53], ionic liquids were used as solvent in the pretreatment and conversion of cellulose. It was shown that cellulose can be dissolved in some hydrophilic ionic liquids, such as 1-butyl-3-methylimidazolium chloride (BMIMCl) [54] and 1-allyl-3-methylimidazolium chloride (AMIMCl) [79].

A patented method for cellulose conversion using ionic liquid [80] includes the following steps:

- contacting the cellulosic raw material with the ionic liquid in the reactor at a temperature below 80°C, preferably below 70°C;
- supply of hydrogen-containing gas to the reactor (pressure 10-50 bar);
- increasing the temperature in the reactor from 80°C to 220°C.

Optimal conditions of the process are as follows: pre-dissolution of cellulose in $ZnCl_2$ x$4H_2$ 0, 5 % Ru/C, hydrogen pressure 40 bar, process temperature 130°C. At a process time of 50 minutes the yield of sorbitol was 88.7 % , at 360 minutes - 97 %. The main disadvantage of the method is the difficulty of separation of ionic liquid and reaction products, as well as a relatively high content of ruthenium in the catalyst.

The authors [57] believe that boric acid can break the crystalline structure of cellulose by forming bonds with a large number of hydroxyl groups, thereby increasing solubility and catalytic activity. Based on this assumption, the authors developed a special additive to ionic liquids.

The compound was synthesized by metathesis between boronic acid and the cation of i-butyl-3-imidazole. While compound (1) is insoluble in water, it is completely miscible with ionic liquids (1- i-butyl-3-methylimidazolium chloride ([BMIMJC1) and trixelyltetradecylphosphonedodecylbenzene sulfonate([TNTBR] - [DBS])).

Without the addition of compound (1), a cellulose conversion of 15% was achieved using [VM1M]C1 nanoclustersJnv [VM1M]C1 as a catalyst [74]. However, using different hydrogen donors and in the presence of additive (1), cellulose was successfully converted to hexitols in high yield. Both ionic liquids [VM1M]C1 and [THTdP] [DBS] were found to be equally efficient when sodium formate was used as a hydrogen source. Furthermore, in the absence of ruthenium nanoparticles and hydrogen, cellulose was hydrolyzed with 95% conversion to glucose using compound (1) for 5 h at 80 °C, proving the crucial importance of compound (1) in the activation of cellulose polymer chains.

Recently, reports have appeared on a specific method of cellulose transformation - hydrogenation with hydrogen transfer, in which 2-propanol (isopropyl alcohol) is used as a hydrogen source instead of H_2 [82]. 2-propanol on the catalyst surface dehydrogenates into acetone and releases two hydrogen ions, which further participate in the hydrogenation of glucose to hexitols (Fig. 1).

At 190°C, after 18 h of reaction on Ru/AC catalyst (Norit) a 43 % yield of hexitols (34 % sorbitol, 9 % mannitol) was achieved. Only some part of the hydrogen ions formed was used for subsequent hydrogenation, most of them were released into the gas phase as H_2 (with a partial pressure of 0.8 MPa) during the initial half-hour period. Given this fact, the authors determined that on the same catalyst the hydrogenation of cellulose was as low at a hydrogen pressure of 0.8 MPa as it was without 2-propanol. Good yields of sugar alcohols (sorbitol 30%, mannitol 8%) were obtained. After the study of Ru/AC catalyst, it was concluded that in the hydrogenation reaction with hydrogen transfer, trivalent or quadrivalent ruthenium showed high activity, while RuO showed no activity.

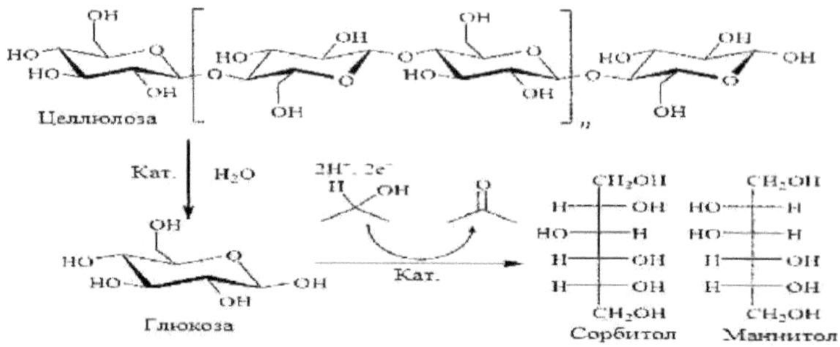

Figure 1 - Hydrogen transfer hydrogenation of cellulose to sugar alcohols [58].

1.2 Enzymatic methods of hydrolytic hydrogenation and hydrolysis of hydrocarbon-containing plant raw materials and wastes

Nowadays, enzymatic hydrolysis is the most popular way to obtain raw materials for the production of cosmetic, chemical and medical products. This process is the hydrolysis of starch-containing materials using enzymes. As a rule, food biomass containing starch is used as a raw material. The use of food biomass as a feedstock increases the cost of the process. In order to avoid the use of this type of raw material, research is being conducted on the enzymatic hydrolysis of cellulosic materials.

The production of D-sorbitol is mainly associated with the hydrogenation of D-glucose and hydrogenolysis of glucose-containing raw materials [59, 60]. The hydrogenation process can be carried out in different ways: biotechnological, electrochemical and using homogeneous and heterogeneous catalytic systems. Of the listed methods, the method that meets a number of requirements is preferable - it is high efficiency, simplicity of technological support, minimization of costs, economic benefit and environmental safety. The catalytic method of D-sorbitol production using heterogeneous catalytic systems, which is currently used in the chemical industry, most of all meets these requirements.

Biotechnological method of D-sorbitol production

Biotechnological processes are based on the functioning of either cells or biological structures isolated from them, most often enzymes. In large-scale production - production of yeast and beverages, ethanol, most organic acids and amino acids, polysaccharides, etc. - fermentation process is realized using active, specially selected cultures of microorganisms. Microorganism strains are also used in the fine biosynthesis of antibiotics, vitamins, and hormones [61]. For example, in the presence of enzyme systems of *Zymomonasmobilis* microorganisms, sorbitol is produced from glucose as a result of a redox reaction [62-64].

In 1984, ViikannBaiTown et al. [65], first discovered the ability of the anaerobic bacterium *Zymomonasmobilis (Z. mobilis)* to synthesize sorbitol from sucrose or from glucose and fructose.

2asyaguyi8sorez[66] investigated in detail the biochemical mechanism of sorbitol synthesis by the bacterium Z. mobilis. They showed that 40000 Dalton glucose-fructose oxireductase (GFO) is a tetromeric enzyme with tightly attached nicotinamidadenine dinucleotide phosphate (NADO). This enzyme is responsible for the processes of fructose reduction to sorbitol and glucose dehydrogenation. In addition, they found the presence of glucono-5-lactonase (GL), which hydrolyzes glucono-8-lactone to gluconic acid (Fig. 2).

Figure 2 Conversion of glucose and fructose to sorbitol and gluconate under the action of GFO and GL

The use of Z. *mobilis* cell strain ATSS-29191 [67] for bioconversion of an equimolar solution of glucose and fructose with concentrations up to 600 g/L allowed us to obtain D-sorbitol and gluconic acid with concentrations of 240 and 210 g/L, respectively. In further experiments, cetyltrimethyltrimethylammonium bromide was used as an optimal emulsifier. Free cells treated with cetyltrimethylammonium bromide achieved 98 - 99% selectivity with D-sorbitol and gluconic acid productivity of 1.8 -2.1 g/($_{gclstoc-h}$). At the same time, it was noted that immobilization of cells on k-carrageenan resulted in the loss of their activity, which was the result of impeded diffusion or inactivation of GFO during immobilization.

In [68], a method is proposed that belongs to genetic engineering and can be used in the microbiological industry. Arabite-producing yeast or fungi are transformed with DNA encoding D- arabitol dehydrogenase forming D-xylose and DNA encoding xylitol dehydrogenase. The transformed yeast or fungi are then grown under conditions that allow synthesis of xylitol. The xylitol is extracted. Yeasts are selected from Lygosaccharomycesrouxii, Candidapolymorpha, Torulopsiscandida, Pichiafarinosa, Torulasporahansenii. Fungi are selected from Dendryphiellasalina and Schizophyllumcommune. The method converts readily available carbon sources such as D-glucose into xylitol.

In [69], cellulose and hemicellulose in the biomass are first decrystallized with concentrated sulfuric acid and subjected to the first hydrolysis, resulting in a hydrolysate saturated with sugar and acid. The silicon oxides contained in the biomass are then removed and sent for recycling. The remaining solid residue is then re-decrystallized and re-hydrolyzed. The resulting sugar solution is then subjected to fermentation using microorganisms, preferably yeasts and bacteria, providing fermentation of both hexose and pentose simultaneously. The method allows the production of sugars from biomass containing cellulose and hemicellulose, which is cost effective and environmentally friendly.

The paper [70] describes a method of pretreatment of cellulose-containing raw materials for enzymatic hydrolysis, which involves preparation of a suspension of raw materials and treatment with acid solution. As an acid solution, nitric acid solution with a concentration of 2.8-6.5% is used. The solution is heated to a temperature of 98-100°C and kept for 1-5 hours. The obtained product is filtered and washed. And miscanthus or fruit shells of cereals are used as cellulose-containing raw material. The acid solution used at the end of one cycle is strengthened with concentrated nitric acid to a solution concentration of 2.8-6.5% and repeatedly used again. The solution used for 18-25 cycles is neutralized with ammonia to obtain ammonium nitrate solution for use as a fertilizer.

A method related to the microbiological as well as food industry and which can be used in the utilization of waste containing cellulose is described in [71]. The method provides pre-treatment of cellulose-containing raw material, of which barley or rice pellet or cotton lint or guza-paya is used to destroy the crystalline structure and increase its specific surface area. This treatment is carried out by extrusion. The obtained extrudate is subjected to enzymatic hydrolysis by enzyme preparations Asp. mix B 2000 or Blmix or celloviridin G20x, at a mass ratio of enzyme: extrudate 1-3:1-9 respectively. The proposed method is waste-free, environmentally friendly and inexpensive, characterized by a high degree of bioconversion of raw materials with a significant reduction in the time required for its implementation.

A method including a stage of hydrolysis of cellulose-containing biomass to obtain an aqueous sugar solution has been developed [72]. The obtained aqueous sugar solution is then filtered through a nanofiltration membrane and/or a reverse osmosis membrane. The purified sugar solution is collected from the inlet side and fermentation inhibiting substances are removed from the filtrate side. Said fermentation inhibiting substances are one or more compounds from organic acids, furan compounds and phenolic compounds. The invention makes it possible to obtain sugar syrup purified from fermentation inhibiting compounds in a simple way and to increase the efficiency of fermentation production of various chemical products.

8yue1ga et al. [73] compared the activity of 6 strains of *Zymomonasmobilis* against D-sorbitol and gluconic acid.

Zymomonasmobilis strain ATCC-29191 showed relatively good growth and high yield of products compared to others. The effect of initial substrate concentration on the yield of D-sorbitol and gluconic acid in a batch process was investigated for this strain [74]. The equimolar concentrations of glucose and fructose ranged from 100 to 750 g/L. Increase in productivity occurred when the initial substrate concentration increased up to 650 g/L. When the initial substrate concentration reached 750 g/l, a decrease in productivity was observed.

It was also found that at an initial concentration of 100 g/L, glucose and fructose were equally converted to sorbitol and gluconic acid, especially during the initial period of the reaction. However, after 20 minutes, a shift in the metabolism of sugars occurred, characterized by: an increase in glucose consumption, a strong decrease in fructose consumption, a decrease in sorbitol yield and a decrease in gluconic acid concentration. Increasing the concentration of substrates up to 600 g/L resulted in almost complete conversion of sugars to sorbitol and gluconic acid.

The increase in product yield with increasing initial concentration of both sugars is a consequence of two effects: loss of cell permeability and inhibition of sugar metabolism. The effect induced by high initial concentrations of substrates was determined by counting the number of cells before the start of the bioconversion process and after the end of the process. The initial cell concentration, at a substrate concentration of 400 g/L, was $2x10^9$ cells/mL and decreased to $2x10^4$ cells/mL after 3 hours of incubation. At an initial substrate concentration of 650 g/L and final cell count after hours, the cell concentration was 8x10 cells/mL. This confirms that an increase in substrate concentration leads to a decrease in cell formation. The experimental results also showed that the resulting products inhibit glucose uptake. Thus, the addition of equimolar concentration of sorbitol and gluconic acid to the cell solution leads to a decrease in glucose consumption and gluconic acid formation [74].

A method for producing xylitol from an aqueous xylose solution is described [75], comprising the steps of fermenting said aqueous xylose solution with a yeast strain capable of converting free xylose into xylitol and free hexoses present in said solution into ethanol. A time sufficient to produce a fermentation solution containing xylitol and to separate one or more xylitol-enriched fractions and said fermentation solution by chromatographic separation has been determined.

The authors [75] carried out enzymatic hydrolysis of the substrate with MultifectXylanase to obtain xylitol

The method of saccharification of lignocellulosic raw materials, which includes preparation of raw materials, including coarse grinding and subsequent fine grinding of raw materials, was developed in [76]. The prepared raw material is treated with an enzyme preparation. As an enzyme preparation used culture liquid obtained by cultivation of strain Penicilliumverruculosum BKM F-3972D. The method allows to increase the yield of digestible sugars. The total amount of sugars in the solution is 6.12-8 kg. The glucose yield is 5.75 kg or 94% of the total sugars. The method [77] of ethanol production includes fermentation, waste sorting, pulverization of the cellulosic component, boiling, and pH correction of the digested mass prior to fermentation followed by ethanol extraction. According to one embodiment, the pulverized cellulosic component (PCI) is treated with concentrated sulfuric acid in a weight ratio of 1:1. The resulting mixture is diluted with water at about 80-100°C. It is boiled with stirring at 100°C and the solid phase is filtered off. The filtrate is separated into a solution containing acid and a solution containing sugar. The sugar is concentrated to 12-14%. The pH correction is made to pH 6. Fermentation of the concentrated sugar solution is carried out using yeast. According to another embodiment, the SCI is subjected to acid hydrolysis.

The hydrolysate is separated by continuous ion-exclusion chromatography into a solution containing acid and a solution containing sugar. The sugar is concentrated to 12-14% by means of a reverse osmosis filter. The pH value is adjusted to 6 with ammonia. Fermentation of the concentrated sugar solution is carried out with yeast at a temperature of 25-36°C to obtain a fermentation, from which ethanol is extracted by distillation. According to a third embodiment, the SCI is hydrolyzed with sulfuric acid at a temperature of 40-100°C. The soluble component is separated from the insoluble component. The insoluble component is dried and treated with concentrated sulfuric acid in a weight ratio of about 1:1 to obtain a partially hydrolyzed mixture. The latter is diluted with water at 80-100°C, boiled under stirring at 100°C, and the solid phase is separated to obtain a filtrate. The filtrate is separated into a solution containing acid and a solution containing sugar. The sugar is concentrated to 12-14%. The pH value is adjusted to 6. Fermentation of the sugar solution is carried out by yeast. The method of obtaining glucose provides for grinding raw materials, treating them with concentrated sulfuric acid at a weight ratio of 1:1 while heating from 30 to 80°C. The resulting mixture is diluted with water at 80100°C, boiled at 100°C with stirring. The insoluble component is then separated and glucose is obtained.

The main disadvantage of this method is the formation of a mixture consisting of sorbitol and gluconic acid, which deteriorates the quality of the target product, in addition, this method is very labor-intensive and not economically viable compared to catalytic methods.

To study the enzymatic hydrolysis of cellulose, pretreatment of cellulose is necessary. In [78], kraft bleached hardwood pulp, cotton microcrystalline cellulose (MCC), and flax fiber were used. All types of pulp used were pretreated to remove mineral compounds from them. The treatment was carried out with 1 n hydrochloric acid solution. Industrial preparation Celloviridin G20x was used for enzymatic hydrolysis process. In addition, we compared hydrolysis using different types of acids (peroxymonosulfuric acid, peracetic acid, and Lewis acid).

Enzymatic hydrolysis was carried out on water swings at 150 rpm in hermetically sealed bins at 55°C. The mass of the sample of cellulose under study was 50 mg, the volume of the reaction mixture was 2 cm^3 , the concentration of the drug under study was taken 0.25 mg/cm. For detailed study of the process, sampling was carried out after 3, 6, 15, 24, 48 h. [78].

The process itself requires the production of preparations, which in turn require significant financial and time costs for their production [78]. In [79], cellulose pretreatment was carried out with the following reagents: alkali, hydrogen peroxide in alkaline medium and their alternating effects. The authors also pretreated with hydrochloric acid to remove lignin. CelloLux-A was used as an enzymatic preparation. The authors showed that the yield of products increased with increasing amount of enzymatic preparation (Fig. 3).

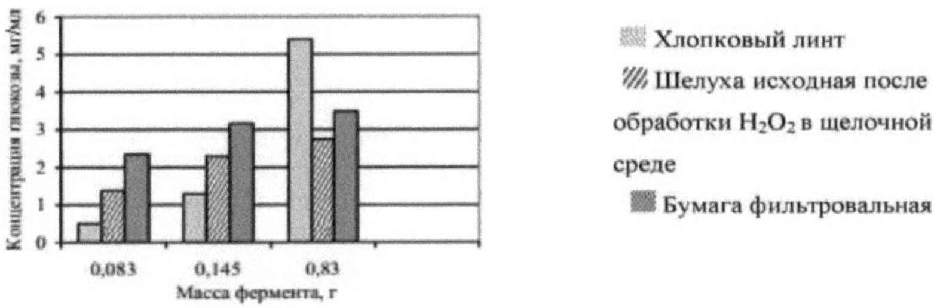

Figure 3 - Dependence of the concentration of glucose formed on the mass of enzyme used

The Japanese company GekkeikanSake announced the successful completion of the development of a technology to produce bioethanol from cellulose using only microorganisms. Until now, the primary breakdown of cellulose to glucose, which is necessary for further fermentation into alcohol, required pretreatment of the feedstock - rice straw and chaff - with chemicals, such as sulfuric acid, or the application of high temperature and pressure. As the researchers report, success was achieved by developing a method for obtaining significant quantities of an enzyme containing chromosomally modified kojimold (aspergilli) bacteria. The resulting microorganisms, capable of breaking down cellulose without additional influences, called Superkojimold. Interestingly, the idea to take a closer look at these bacteria came because they are used in the traditional sake production process. According to the company, the method developed by the company saves the energy needed to break down cellulose and distill bioethanol after fermentation, and also reduces the amount of water needed in the technological cycle [80].

Analyzing the above, it should be noted that for the use of enzymatic hydrolysis on an industrial scale, there is a problem of alkali and acid utilization. To solve this problem, additional financial costs are required, which will affect the cost of the final product.

In addition, the cellulosic material needs to be further treated to extract lignin from it. For this treatment hydrochloric acid is used, which in turn must be disposed of. The use of acids and alkalis leads to rapid wear and tear of the equipment. The use of enzymatic preparations increases the cost of the final product. This is due to the fact that additional financial costs are required to produce this type of preparation. The use of enzymatic hydrolysis of cellulose on an industrial scale is not feasible.

1.3 Catalysts for chemical and enzymatic hydrolytic hydrogenation and hydrolysis of hydrocarbon-containing plant raw materials and wastes

Heterogeneous catalyst has certain advantages from a practical point of view compared to homogeneous catalyst: it can be easily separated from the products, is thermally stable and can be regenerated.

The most widely used heterogeneous catalysts for the hydrogenation of sugars are platinum metal and nickel-based catalysts.

The catalytic hydrogenation reaction of D-glucose to D-sorbitol, which is thermodynamically spontaneous and exothermic (K_{42} h k = 473), proceeds according to the following equation:

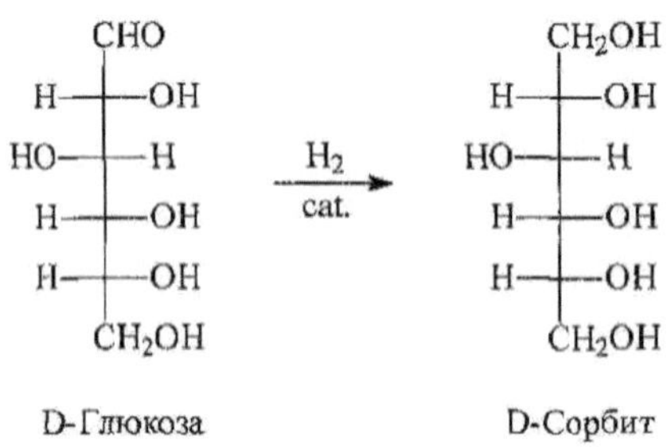

As industrial catalysts for this process, first of all, it is worth mentioning Ni-Rhenium and skeletal nickel, one of the most active catalysts obtained from aluminum-nickel alloy by dissolving a part of aluminum in alkali. For example, the reaction of hydrogenation of glucose in the presence of skeletal nickel is carried out at a hydrogen pressure of 8 MPa and a temperature of about 140°C. The process is not selective and with a decrease in the catalytic activity of the catalyst.

The hydrogenation of D-glucose to D-sorbitol on Ni/kieselguhr was studied by Tukac[81]. Studying the hydrogenation process of 40% aqueous solution of glucose in the temperature range from 115 to 165°C and hydrogen pressure in the range from 0.5 to 10 MPa, the author found the hydrogen partial order, which was 0.65, and calculated the value of apparent activation energy 23.8 - 48.5 kJ/mol. This value indicates that the reaction depends on external diffusion factors, primarily due to the high initial molar concentration of glucose and the viscosity of its solution.

Besides using nickel/kieselguhr as a carrier, other substrates such as SiC>2, TiO_2 , A1 O_{23} [82] and coal can also be used. Catalyst #/A1 0_{23} with different metal percentage, between 15 and 30%, was synthesized using three methods, namely homogeneous leaching (HA), double hydroxide (LC) and citrate (CP) methods [83].

The authors [82] found that Ni/Al 0_{23} - 2 and Ni/Al 0_{23} - 5 catalysts have the highest activity compared to commercial Ni/kieselguhr. These catalysts have high activity (over 90 mmol/g Ni/h) and are able to carry out the hydrogenation process with high selectivity (~80%).

The influence of methods of preparation of nickel catalysts (deposition, impregnation, sol-gel method, as well as matrix synthesis) and different substrates (SiC>2, TiO_2 , A1 O_{23} , coal) was studied in [84]. The authors [83] experimentally found that the substrate activity decreases in the series A1 0_{23} >TiO_2 >$Si0_2$ > C. During hydrogenation of 50% glucose solution in an autoclave at 120°C and 12 MPa hydrogen pressure using a catalyst obtained by the sol-gel method, it was found that the catalyst synthesized by a similar method showed no catalytic activity in the hydrogenation of glucose.

Although different substrates were used for the active metal, only one type of catalyst was used for this synthesis [85, 86]. The preparation of the catalyst was carried out in one step. A suspension of the precursor (nickel ethylacetonate) in a matrix solution was used.

For a series of catalysts (Ni_2 OOrgl, Ni $OOrg_{22}$ H Ni $OOrg_{23}$), an increase in pore diameter with increasing toluene concentration was expected, but the opposite effect was also found. It is related to the competition between nickel acetylacetonate solutions and toluene for space in the micelle during the preparation of the catalyst. This was the reason for the decrease of nickel content in the catalyst with increasing toluene concentration.

In kinetic testing, most of the catalysts obtained with ethylenediamine provided good yields of sorbitol. Ni_2 OOrgl catalyst with high metal content and wide pores provided the highest yield of sorbitol (7.7%). No.$_2$ 08 with low nickel content and narrow pores was characterized by low sorbitol yield (0.5%). The authors suggested that the low conversion is a consequence of mass transfer in long and relatively narrow pores. Catalyst №=08p1 showed, despite the high content of nickel and wide pores, providing high mass transfer and availability of the substrate for catalytically active centers, very low yield of sorbitol in the process of hydrogenation of glucose, which is probably due to its insufficient reduction. It was also found that in the presence of different catalysts, the most reactive form of glucose was the open form, so that the high concentration of this form contributed to the high conversion of glucose, while the choice of substrate significantly affected the equilibria between the different tautomeric forms of glucose.

The invention [87], which relates to a catalyst for hydrolysis of cellulose or hydrolysis of cellulose and recovery of hydrolysis products and a method for producing sugar alcohols from cellulose, is quite interesting. A catalyst for hydrolysis of cellulose and hydrolysis of cellulose and recovery of hydrolysis products, in which a transition metal of groups 8-11 is deposited on a solid substrate, is described. A method for producing sugar alcohols is also described, comprising: hydrolysis of cellulose in the presence of a catalyst in a hydrogen-containing atmosphere at elevated pressure, and recovery of a cellulose hydrolysis product. The technical effect is easy separation of the catalyst from the product, no need for pH adjustment, neutralization of acid or alkali in obtaining alcohols, possibility of reuse of the catalyst without its activation, obtaining sugar alcohols directly from cellulose using such a catalyst.

The authors [88] proposed a method comprising a hydrolysis process using an acid cluster catalyst in a pseudo-molten state to hydrolyze cellulose contained in plant fibrous material and produce glucose. During the hydrolysis process, the cluster acid catalyst and the first amount of plant fiber material increases the viscosity of the cluster acid catalyst. When added to the cluster acid catalyst in a pseudo-molten state, the cluster acid catalyst and the second amount of plant fiber material are heated and mixed. When the viscosity of the hot mixture of the cluster acid catalyst and the first amount of plant fiber material is reduced, homopolyacids or heteropolyacids are used as the cluster acid catalyst. The proposed method allows increasing the amount of treated plant fiber material per unit mass of cluster acid catalyst.

N. DechampH et al. [89] studied the kinetics of catalytic hydrogenation of glucose. A commercial nickel catalyst on silicon-aluminum substrate consisting mainly of 48.4% Ni, 5.15% A1, 8.46% Si, 0.27% Mg, 0.34% Na, and 0.47% Fe was used as a catalytic system. The authors studied in detail the effects of external and intra-diffusion factors on the hydrogenation process. When studying the kinetics of hydrogenation using this catalyst, an apparent activation energy value of 67 kJ/mol was found. Its value is much higher than the apparent activation energy in the diffusion region (1221 kJ/mol), indicating that the reaction occurring on the surface of the catalyst occurs in the kinetic region. Kinetic modeling from experimental data showed that the reaction occurs between glucose and hydrogen adsorbed on the nickel surface. Such a course of the reaction indicates the reaction mechanism, which proceeds in accordance with the Langmuir-Hinshelwood model.

Nickel is an economically inexpensive metal, and catalysts based on it, such as Ni-Reneia, are historically considered to be highly active [90]. To increase the activity of nickel catalysts, they are often promoted by the addition of mainly inorganic elements (B [91], P [92] and various metals). N. Lrn et al. [91] used an alloy of nickel and boron promoted by chromium, molybdenum and tungsten. The catalyst #-B/8Yu$_2$ was chemically reduced by KBH_4 in aqueous solution. Its amorphous structure was confirmed using the following methods: XIV(X-ray diffraction method) and EXAFS(extended fine structure in X-ray absorption spectra method). In the hydrogenation of glucose in aqueous phase, the pre-prepared amorphous catalyst Ni-B/ZU$_2$ exhibited much higher activity than other nickel catalysts such as crystalline Ni-

B/SiO_2 , Ni-B/SiO_2 reduced by hydrogen, as well as commercial Ni-Reneia. This was due to the combined influence of the catalyst structure and the electronic effect between metal and boron. The Ni-B/SiO_2 preconditioning influenced its activity, the maximum catalyst activity was observed by Sid_2 H calcination of the precursor at 200°C for 2 hours. The presence of promoters (Cr, Mo or W) also increased the activity of the catalyst, they influenced the electronic structure of the active metal surface, which had a favorable effect on the adsorption and polarization of the carbonyl group (C=0) of the glucose molecule. Tungsten had the best promoting effect. P. Gallezoto et al. [93] studied Ni-Reneia promoted by molybdenum, chromium, iron and tin. However, nickel catalysts occupy a disadvantageous position in relation to other catalysts for the hydrogenation of sugars, due to the leaching of nickel, resulting in a loss of catalyst activity, as well as an increase in the concentration of nickel in the sorbitol solution. For the use of sorbitol in the food, medical and cosmetic industries, nickel, being a heavy metal, must be completely removed from the sorbitol solution, which entails an increase in economic costs. In addition, catalysts on other active metals such as cobalt, platinum, palladium, rhodium and ruthenium are more preferable [92, 94, 95].

The authors [96] showed that the conversion of cellulose in polar solvents is very problematic. This problem is solved by the use of Pt catalyst, which uses its pores to facilitate the process of interaction between the catalyst and cellulose, which favors the process of hydrolytic hydrogenation of cellulose

The conversion [97] of lignocellulose is a crucial topic in the renewable and sustainable chemical industry. However, the separation of cellulose from lignocellulose is a task not solvable in polar solvents and therefore difficult to convert into value-added chemicals. A strategy to overcome this drawback is the use of mesoporous carbon, which increases the proximity between the cellulose and the catalyst through its rich functional groups and large homogeneous pores. Here, we report regarding the preparation of a Pt catalyst supported on a type of 3D mesoporous carbon treated with Echinometramathae (Pt/CNE) to increase the interaction between the catalyst and the insoluble reagent. In the hydrolytic hydrogenation of cellulose, the rich oxygen groups of CNE facilitated the access of cellulose to the catalyst surface, and the open pore structure helps oligomers to transform and efficiently diffuse to the active sites in the pore. The highly porous Pt fulfills the dual role of hydrolysis by protons forming hydrogen or aqueous, and efficient hydrogenation. The use of the Pt/CNE catalyst resulted in approximately 80% yield of xylitol, reported the best work to date. In direct polysaccharide conversion on Pt/CNE shows good performance in sugar alcohol production (23% yield). They hypothesize that the openly structured 3D carbon will be widely applied to the conversion of various lignocellulosic materials.

The authors [98] carried out hydrolysis and hydrogenation of cellobiose in the solvent ZnC H_{1242} O In order to optimize the synthesis of D-sorbitol, the following conditions were optimal: Ru/C-catalyst, temperature 125°C, solvent ZnC H_{122} O, sorbitol yield 95%.

The hydrogenation catalysts commonly used in the literature are carried out on metallic catalysts such as ruthenium on carbon, ruthenium on carbon nanotubes, ruthenium modified zeolites, nickel on carbon nanofibers, tungsten carbide on carbon nanotubes; ruthenium nanoparticles and modified nickel catalysts. They have been successfully used as catalysts for the conversion of cellulose to polyols. In this study, Ru/C catalysts were selected for their high efficiency and no leaching of the catalysts was observed in the hydrogenation of glucose [99, 100-110]

2 PREPARATION OF MATERIALS, METHODS OF ANALYSIS AND RESEARCH

2.1 Characterization of research objects

According to the results of research in 2015, this paper investigated beer pellet formed as waste at the plant "Shymkentpivo". Malt pellet (OST 18-341-79 "Raw beer pellet") is formed as a residue after separation after the liquid phase - beer wort in the process of mash filtration. The shot consists of liquid (45%) and solid (55%) phases. The solid phase contains the shell and the insoluble part of the grain. Raw beer grist is a light brown colored grist with a specific odor and taste. It can contain up to 88% water and can be stored for 24 hours at ambient temperature. The chemical composition of pellets varies depending on the quality and assortment of processed grain products and the type of beer produced (Table 1).

Table 1- Chemical composition of beer pellets

Name of components	Content, %
Ash substances	5,50
Easily hydrolyzable polysaccharides	21,32
Difficult-to-hydrolyze polysaccharides	24,66
Gecosans	17,68
Pentosans (without uronic acids)	28,03

As the second raw material, we studied cellulose obtained from guza-paya by explosive autohydrolysis. The chemical composition of cellulose is given in Table 2.

Preliminary investigated plant raw materials were crushed and sorted. Raw materials fractionated through sieves with particle size 2-3 mm were used for chemical analyses. Chemical analyses were carried out using the following methods:

- humidity - by drying a sample of raw materials in a desiccator at 105°C to constant weight;
- ash substances - by combustion of a sample of raw materials followed by

calcination in a muffle furnace at a temperature of 600°C ;

- the content of easily- and hard-hydrolyzable polysaccharides was determined according to the method of Kiesel and Semiganovsky;
- lignin - by the Koenig method modified by Komarov using 72% sulfuric acid;
- volume mass of raw materials was determined by weighing a certain volume of plant raw materials on analytical scales;
- sugars were analyzed by the Bertrand and Macken-Schoorl method
- individual sugars were determined by HPLC liquid chromatograph; ShimadzuLClO-ATVP, TEST- 900 differential digital detector, Luna 5 columns.

Table 2 - Chemical composition of cellulose from guza-pai

Characteristics m.d., %* naa.s.s.	
Cellulose	88,55
Moisture, %	6,2
Ash	4,75
Pentosans	7,89
Lignin	10,82

2.2 Methodology for determining moisture content

Weights of raw materials not more than 3 g are placed in pre-washed, dried and weighed glass bouquets. Weighing of the weights in the bins is carried out together with the lids.

Weighed bins with suspensions are opened and placed together with the lids in a drying oven preheated to 105-110° C and dried at this temperature for 2 hr. Burets removed from the cabinet are closed with lids, placed on a ceramic tile or metal sheet and cooled for at least 5 min, and then in the desiccator to room temperature and weighed. All weighings are carried out with an error of not more than 0.0002 g.

Humidity *w is* determined by the formula:

$$w = \frac{m_{вл}}{m_{нав}} = \frac{m_{нав} - m}{m_{нав}}, \tag{1}$$

where *tel* - mass of moisture, g;

tnav - mass of the sample before drying, g;

m - mass of the sample after drying, g.

The mass of the sample before drying is determined by the formula:

$$m_{нав} = m_o - m_m \tag{2}$$

where *to* is the mass of the crucible with the suspension, g; *tt* is the mass of the crucible, g.

Moisture is determined in parallel in three samples. The final result is the arithmetic mean of three parallel determinations within the tolerance of differences in moisture content, which should not exceed 0.5%. Calculation of the results of the analysis is made to the second decimal place. The final results are rounded to the first decimal place.

It is allowed to use metal cups with a lid with a diameter of 50-60 mm and height of 30-40 mm, the pulp suspension is dried at a temperature of 130-140 C for 30 minutes, cooling in the desiccator without a thermometer [36].

2.3 Methodology for determination of the mass fraction of pentosans

The cellulose suspension is placed in a round bottom flask. Pour 20 g of sodium chloride into the same flask and use a measuring cylinder to pour 100 cm of^3 solution with a mass fraction of hydrochloric acid 13 % (Fig. 4).

The round bottom flask is placed in a glycerol bath and connected to the unit. The level of glycerol should be kept 1.5-2.0 cm above the level of liquid in the flask. A thermometer is fixed in the glycerine bath. The unit is placed in a fume cupboard.

The distillate should be collected in a measuring beaker or cylinder placed for cooling in a crystallizer with cold water.

The heating of the glycerin bath is adjustable. The temperature of the glycerol should be maintained at 164-166° C. The distillate should be distilled off at a rate of 30 cm³ per 10 min. After distillation of each 30 cm³ from the drip funnel 30 cm³ of a solution with a mass fraction of hydrochloric acid 13 % are poured into the round bottom flask.

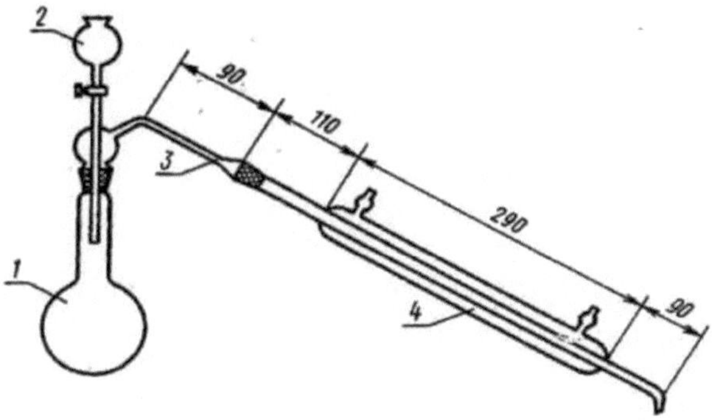

Figure 4 - Installation for the determination of the mass fraction of pentosans: 1 - round bottom flask with a capacity of 500 cm³ ; 2 - dropping funnel; 3 - tube; 4 - refrigerator.

Distillate 225 cm of³ distillate.

After the end of distillation the distillate from the measuring beaker is transferred to a measuring flask with a capacity of 250 cm³ , the beaker is rinsed with a small amount of solution with a mass fraction of hydrochloric acid 13 % and the volume in the flask is brought to the mark with a solution of the same acid.

The flask is corked and the contents in the flask are mixed thoroughly.

After stirring take 5 cm of³ solution by pipette and transfer it to a 100 cm flask³ . In the same pipette with the help of a rubber pear, 25 cm³ of the solution of orsine is added, then the flask is corked and the contents of the flask are stirred and kept for 50 min at (20±0.5)° C.

After 50 min add 20 cm³ of ethyl alcohol to the contents of the flask by pipette. After stirring the contents of the flask are slightly cooled and kept at (20±0.5) C for 10-15 min.

At the same time a control sample is prepared in the same way, in which the analyzed solution is replaced with 5 cm³ of a solution with a mass fraction of hydrochloric acid

13 %. The control sample is kept together with the test samples.

In 10-15 min after preparation of solutions determine their optical density relative to the control sample on photoelectrocolorimeter with a wavelength of 630 nm (red filter).

Preliminary selection of cuvettes is made depending on the value of the optical density of the test sample. The nominal value of optical density should be within the range 0.3-0.5.

Cuvettes with the thickness of light absorbing layer 10, 20 and 50 mm are used.

According to the obtained value of optical density, the mass of pentosans in the sample is found using the appropriate calibration graph.

Mass fraction of pentosans as a percentage of absolutely dry cellulose is calculated according to the formula:

$$\Pi = \frac{g \times 100 \times 100}{m \times (100 - W) \times 1000} = \frac{g \times 10}{m \times (100 - W)}, \qquad (3)$$

where g is the mass of pentosans in the sample, mg;

m - weight of air-dry pulp sample, g;

W - pulp moisture, %

For the result of the analysis should be taken as the arithmetic mean of the results of two parallel determinations, rounded to tenths of a percent. The discrepancy between parallel determinations at a confidence level P = 0.95 should not exceed 0.2 % for the mass fraction of pentosans up to 3 % and 0.4 % for the mass fraction of pentosans over 3 % [33].

2.4 Determination of cellulose content by weight method

From the combined sample of air-dry cellulose, cut into pieces of size 10×10 mm, take a suspended weight of about 3 g, weighed with an error of no more than 0.0002 g, put into a porcelain beaker with a capacity of 150 cm^3 and pour 45 cm^3 of 17.5% NaOH solution, the temperature of which is (20±0.2)° C. In a separate sample determine the moisture content of cellulose. Sodium hydroxide solution is poured in portions: at first

15 cm of³ is poured in and the cellulose is stirred gently for 2-3 minutes with a glass rod with a flat end. Then add the rest of the solution (30 cm^3) and stir evenly and gently for 1 min. The beaker with the mixture is covered with an hour glass and placed in the thermostat with temperature (20±0.2)° C for 45 min, counting from the beginning of treatment of cellulose with alkali.

After this time 45 cm^3 of distilled water (20±0.2)° C is added to the mass, gently stirred for 1.5 min and transferred to a glass filter. The pulp mass is evenly distributed on the filter and the filtrate is sucked off into a receiving flask. During filtration to avoid fiber losses the filtrate is passed twice through the fiber layers. The residue on the filter under a weak vacuum is washed in three portions of 25 cm^3 with 9,5 % NaOH solution with temperature (20±0,2) T . Each new portion of washing alkali is added only after complete suction of the previous portion. Total duration of washing with lye should be 2-3 min. After suction of alkali the fiber is washed with separate portions of distilled water at temperature 18-20 T with intermediate suction. Washing is carried out to neutral reaction according to phenolphthalein. At the end of washing the suction is continued until the drops disappear at the tip of the funnel when compacted with a stick. The washed residue (alpha-cellulose) is transferred with tweezers into a previously dried to a constant mass bouquet and dried in a desiccator at (103±2) T for 6-7 hours to a constant mass. At the end of drying the box is placed in the desiccator, cooled and weighed.

Mass fraction of cellulose, % to absolutely dry cellulose, is calculated by the formula:

$$X_{\alpha} = \frac{m_1 - m}{g} \cdot 100, \tag{4}$$

where m is the mass of the empty box, g;

ml - mass of the box with dried alpha-cellulose, g;

g - absolutely dry weight of cellulose, g.

The arithmetic mean of two parallel determinations rounded to 0.1 % shall be taken as the test result. The discrepancy between two parallel determinations shall not exceed 0.3 % [37].

After cooling the reaction mixture, the composite and the residue of unhydrolyzed cellulose were separated by filtration. A pleated filter made of an ashed filter paper was used for filtration. After separation of the solid phase, the volume of the reaction mixture was brought to 100 mL in a glass measuring flask and analyzed by HPLC.

The analysis was carried out using an UltiMate 3000 high-performance liquid chromatograph (Dionex, USA) equipped with a refractometric detector. A Reprogel-H column (500 × 10 mm, number of theoretical plates - 160,000) was used to separate the reaction mixture into fractions. The temperature of the column was 25°C. Sulfuric acid solution (9 mmol/L) with a pressure of 6.5 MPa and a flow rate of 0.5 mL/min was used as eluent. To obtain more accurate result, the sample was taken twice for analysis. The analysis time for one sample was 60 minutes.

2.5 Hardware design

2.5.1Methodology of hydrolytic hydrogenation

Hydrolysis of polysaccharides of plant raw materials and hydrolytic hydrogenation of the formed polysaccharides under hydrogen pressure were carried out in a reactor (1) with intensive stirring, volume 100 cm^3 made of stainless steel with hermetic drive (2) (Fig. 5). The rotation speed of the screw stirrer (3) (1400 rpm) provided the maximum surface of the gas-liquid interface. To carry out the hydrogenation reaction, accurately weighed on analytical scales amount of starting substance, catalyst and the required amount of acid solution is introduced into the reactor (1) by means of a tube (4). The sealed reactor is then purged with hydrogen to remove air. The required hydrogen pressure is set, controlled by a reference manometer (5), and at the same time the electric drive (2) for intensive stirring and thermostat (6) for heating the reactor are switched on. The time is counted from the moment of connection of the thermostat (6) and electric drive.

Electric hydrogen is supplied from the cylinder (7) by means of a reducer.

After the end of the experiment the reactor is cooled, hydrogen is vented and the catalyst is drained through the connection (8).

At the end of the experiment, the catalyst and the unhydrolyzed polysaccharide were separated by filtration. The filtered catalysts were analyzed for the content of reducing agents by the McEuen-Schoorl method. In addition, the pH of the catalyst was measured using a universal ionometer EV-74. After that water was distilled off by vacuum distillation on a rotary-vacuum evaporator. The obtained product in the form of syrup was used for chromatographic analysis.

The developed improved small-sized laboratory universal installation for the study of the process of chemical hydrolytic hydrogenation of beer pellets (Fig. b) allows to vary the temperature from 50° C to 200° C. The loading of beer pellets and acid and reducing agent solution is carried out through the neck 2 welded to the upper part of the body 1. Loading of beer pellets, acid and reducing agent solution is carried out through the neck 2 welded to the upper part of the body 1.

After simultaneous loading of all components, the resulting suspension of hydrolyzable material is quickly heated to the set temperature by the "external" heater 5, which is a ceramic insulated coil wound around the apparatus body. When the set temperature is reached, the "external" heater is switched off and the automatic thermoregulator is switched on, supplying voltage to the built-in cartridge-type heater 6. During the hydrolysis process, sampling is carried out through a mesh filter 7, installed at the level in the middle of the liquid phase column, and a tube-in-tube cooler 8.

When the filter 7 is dirty, it is cleaned by short-term supply of compressed air to the sampling device. To prevent dangerous overpressure, for example, in case of thermoregulator failure, there is a safety valve 3 on the hydrolyzer body, set to the limit pressure of 0.43 MPa.

A typical experiment on hydrolysis-hydrogenation of cellulose was carried out as follows. Suspensions of 50 mg of cellulose and 50 mg of the applied catalyst were placed in the reactor, then 45 ml of isopropanol was added. The reactor was closed, purged with argon (gas pressure 1MPa) and heated to 180 °C. After reaching the set temperature, the reaction was started. Aliquots of ~1 mL were taken from the closed autoclave at 0, 1, 2, 3, 5, and 7 hours for analysis through a special sampler during the course of the experiment.

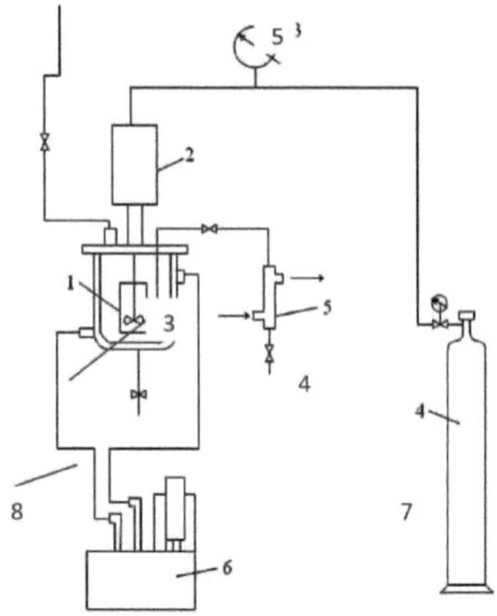

Figure 5 - Schematic diagram of the unit for hydrolytic hydrolysis and hydrogenation of guza-pai cellulose

1 - reactor, 2 - electric drive, 3 - stirrer, 4 - tube, 5 - pressure gauge, 6 - thermostat, 7 - hydrogen cylinder, 8 - connector

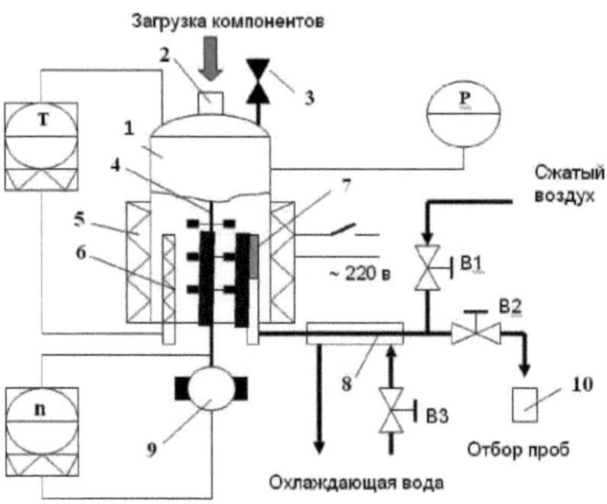

1 - 100 ml reactor vessel

B1 - Filter purge valve

2 - Component loading neck

3 - Safety relief valve

4 - Frame agitator

5 - Electric heater 2.0 kW

6 - Electric cartridge type heater 0.5 kW

7 - Strainer

8 - Tube-in-tube refrigerator

9 - Stirrer drive, 300 watt

10 - Sampling Tank.

B2 - Sampling valve

B3- Cooling water supply valve

P - Indicating manometer MPTS-100, 4 atm, cl.1.5

T - Thermoregulator OVEN TRM1 with

temperature sensor TSP 100 n - Stirrer speed controller with power consumption display.

Figure 6- Schematic diagram of a chemical hydrolytic hydrogenation unit for beer pellets

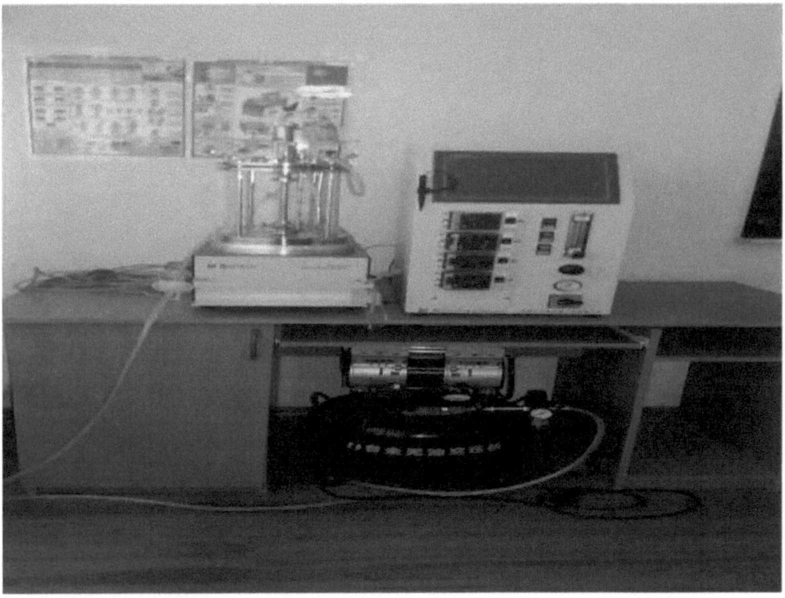

Figure 7 - General view of the installation for the study of enzymatic hydrolysis processes

A fermenter was used to study complex biotechnological processes.

Such installation as a part of the complex allows to carry out research of enzymatic hydrolysis processes of cellulose-containing raw materials at hydromodule 1:10 and more and regulated temperature in the range of 30°C - 75°C.

The drive of the stirrer of the fermentolyser is identical to that of the stirrer of the hydrolyzer and provides power metering at the motor armature.

The general scheme of the unit for fermentolysis does not differ significantly from the standard fermentation unit equipped with a reactor with a mechanical stirrer. Acidity is controlled by pH-meter MULTITEST IPL-513, titration is carried out according to the variant of one-way regulation with titrant supply by peristaltic pump-doser.

In the experiments, beer pellet was used as a feedstock (Fig. 7).

3.1 Development of technology of chemical hydrolytic hydrolysis and hydrogenation of hydrocarbon-containing plant raw materials and wastes based on the use of a combined (hybrid) method

3.1.1Development of technology for chemical hydrolytic hydrolysis and hydrogenation of guza-pai cellulose based on the use of a combined (hybrid) method

In the research work carried out in 2015, we investigated the process of autohydrolysis and chemical hydrolysis of hydrocarbon-containing plant raw materials - guza-paya to obtain cellulose and glucose. To obtain sorbitol, according to the plan of 2016, we separately studied the process of sorbitol synthesis from glucose and developed a combined (hybrid) process of hydrolytic hydrolysis and hydrogenation of cellulose guza-paya. For realization of the first process the most widely used heterogeneous catalysts for hydrogenation of sugars are catalysts based on platinum metals and nickel. The catalytic reaction of hydrogenation of glucose into sorbitol, which is thermodynamically spontaneous and exothermic, occurs according to the following equation:

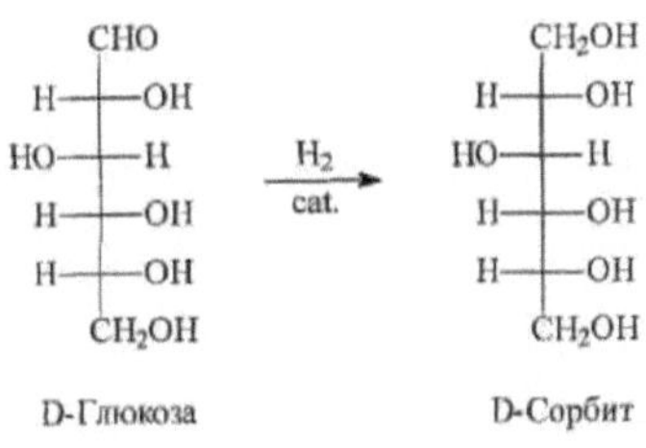

According to this pathway, we must first carry out the process of hydrolysis of guza-paia to produce glucose according to this reaction

This process is very long and involves the formation of many other compounds, and numerous purification steps and the isolation of intermediates are required.

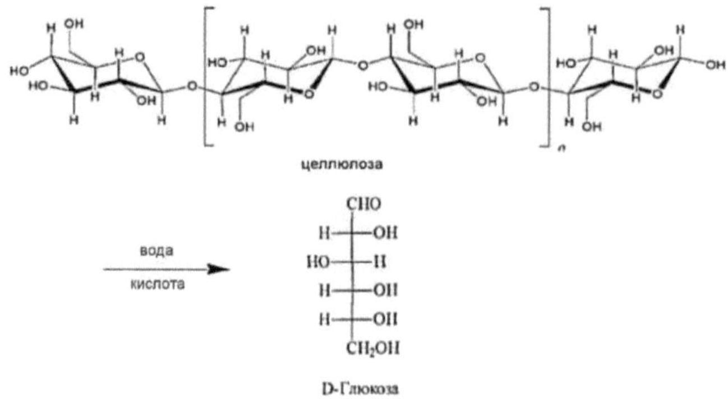

To solve the above problem, we carried out a reaction to reduce the molecular weight of cellulose using a solid catalyst by combining the hydrolysis and hydrogenation processes. The study led to the fact that we carried out the combination of hydrolysis and hydrogenation reactions of cellulose guza-pai in water with the help of a metal catalyst under hydrogen overpressure. In this case, as the following reaction equation shows, the polyatomic alcohols (sorbitol and mannitol) are synthesized in one step.

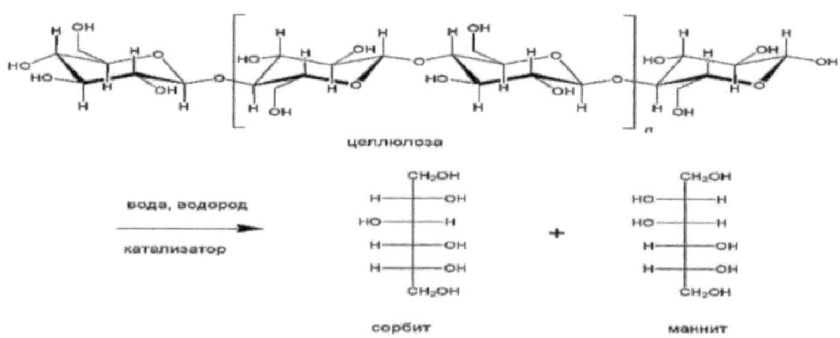

Cellulose guza-pai obtained by explosive hydrolysis method was used for realization of combined (hybrid) or hydrolysis-hydrogenation process of sorbitol production. The process of chemical hydrolytic hydrogenation of cellulose guza-pai was carried out in a steel reactor with a volume of 100 cm^3 in aqueous medium under intensive stirring in the temperature range of 140-220° C, hydrogen pressure of 2.010.0 MPa and duration of the reaction-20-100 minutes. The applied nickel NiZAl O_{23} catalyst was prepared by the impregnation method, and in the process of preparation it was additionally added ferroalloy (FS) in the amount of 5% of nickel weight. To carry out the hydrogenation-hydrolysis process, an accurately weighed on analytical scales amount of cellulose, catalyst and the required amount of water is placed in a reactor (1). The sealed reactor is then purged with hydrogen to remove air. At intervals, samples were taken using a sampler to analyze for sugar alcohols and to determine the degree of cellulose conversion. Cellulose conversion was calculated as the ratio of the weight of unhydrolyzed cellulose to its initial amount. Concentrations of the main products of the process: xylitol, mannitol, xylose, glucose, sorbitol, etc. were determined in the liquid phase of the catalyst by liquid chromatography. Based on the obtained values, the selectivity value was calculated as the ratio of the mass of the formed product to the mass of the hydrolyzed polysaccharide.

The efficiency of the process was determined by cellulose conversion, selectivity and yield of main products.

Table 3 and Figure 8 summarize the experimental data on the investigation of the patterns of change in the rate of chemical hydrolytic hydrolysis and hydrogenation of cellulose guza-pai from the reaction time. The reaction time was varied from 20 to 100 minutes. The optimum time of the process of catalytic conversion of cellulose guza-pai under the conditions chosen by us was determined to be 60 minutes.

Table 3 - Dependence of the rate of chemical hydrolytic hydrogenation of cellulose guza-pai on the time of the process.Experiment conditions: 0.5 g. 3% Ni/Al_2O_3(ΦS), Top= 180° C, P¾=6.0 MPa

№/№	τ, min	Conversion rate,%	Sorbitol selectivity,%	Mannitol selectivity,%	Total yield,%
1.	20	38,2	15,1	3,0	18,9
2.	40	44,5	17,7	3,1	21,3
3.	60	53,0	21,8	3,0	22,6
4.	80	54,4	20,3	2,2	23,7
5.	100	55,0	20,0	2,1	23,9

Up to the sixtieth minute the cellulose conversion reaction is negligible, and after sixty

its values are within error. The same pattern is observed with the sorbitol selectivity index.

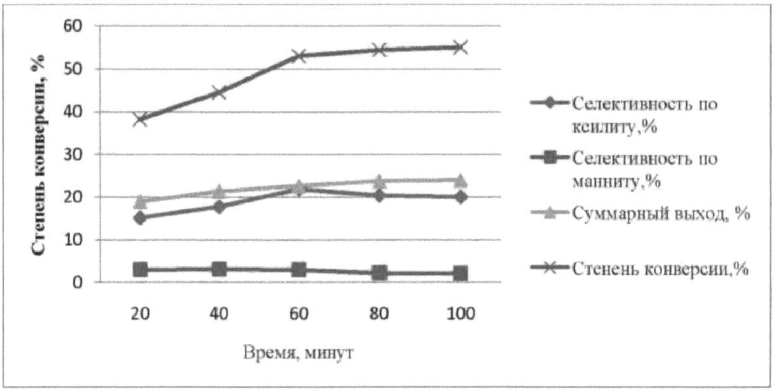

Figure 8 - Kinetic dependences of the process of hydrolytic hydrolysis and hydrogenation of guza-pai cellulose

3.1.2. Development of the technology of chemical hydrolytic hydrolysis and hydrogenation of beer pellets based on the use of a combined (hybrid) method

3.1.2.1. Investigation of combined (hybrid) process of chemical hydrolytic hydrolysis of beer pellet hydrogenation

It is known that most chemical industrial processes require the use of numerous reagents and several reactors. Single-reactor processes, the main difference of which is the use of one reactor for several consecutive reactions without separation of intermediate products, allow to reduce costs dramatically.

We have developed a single-reactor technology for xylitol production from beer pellets, which leads to the absence of processes for separation and purification of intermediate products. As a result of using the combined method, the overall speed of product production increases with less reagents used.

At realization of the first stage we used the following acids - sulfuric, nitric, phosphoric and sulfuric acids, which carried out the process of breaking some chemical bonds in polysaccharides of beer mash. In this process an intermediate product -xylose - is formed, as well as a large number of by-products, including aromatic compounds. In terms of activity, the acids used in this process form the following series:

$$H_2SO_4 > HNO_3 > H_2SO_3 > H_3PO_4$$

The ratio of the rates of hydrolysis of polysaccharides in their presence is as follows:

$$W_{H2SO4} : W_{HNO3} : W_{H2SO3} : W_{H3PO4} = 3{,}1 : 2{,}5 : 2{,}3 : 1{,}6$$

It should be noted that when the reaction is carried out in hydrochloric acid, when a certain concentration of monosaccharides is reached, the samples change color with the appearance of a strong unpleasant odor, which indicates the formation of by-products.In our opinion, by-products are formed as a result of the decomposition reaction. Decomposition reaction of monosaccharides under the action of acid and temperature is irreversible. This reaction consists in dehydration of monos to anhydrides, from which levulinic and gluconic acids and coloring substances are formed as a result of further transformations.

The second stage involves the reduction of xylose to xylitol in isopropyl alcohol solution in the presence of nickel, copper or cobalt catalysts promoted by ferroalloys.

As a result, a maximum 75% yield of xylite was achieved after 2-hour reduction at a reactor temperature of 125 °C. In terms of activity, the developed catalysts form the following series: Ni-FS>Co-FS>Cu-FS. In addition, it is important that the use of pressurized hydrogen was avoided under the developed conditions.

Table 4 and Figure 9 summarize the experimental data on the investigation of the patterns of change in the rate of change of the process of chemical hydrolytic hydrolysis and hydrogenation of beer pellets from the reaction time. The reaction time varied from 40 to 200 minutes

The optimum time of the process of catalytic conversion of beer pellets under our chosen conditions was determined to be 120 minutes. Up to this minute of reaction, the conversion of beer pellets smoothly increases, and after this indicator its values are within the error. The same pattern is observed with the selectivity index for xylitol.

Table 4 - Dependence of the rate of chemical hydrolytic hydrolysis and hydrogenation of beer pellets on the process time

№/№	τ, min	Conversion rate,%	Xylitol selectivity,%	Mannitol selectivity,%	Total yield,%
1.	40	16,4	8,5	1,2	8,8
2.	80	48,7	16,9	2,5	20,0
3.	120	75,0	19,8	2,1	21,8
4.	160	72,8	19,0	2,1	20,6
5.	200	63,3	17,1	2,1	19,9

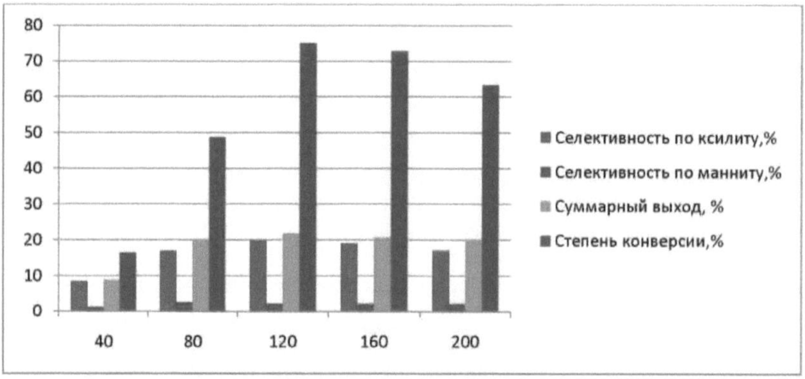

Figure 9 - Experiment conditions: 0.5 g. 3% Ni/Al_2O_3(ΦS), Top= 180° C, P_{H_2} =6.0 MPa

3.2 Development of the technology of enzymatic hydrolytic hydrolysis and hydrogenation of carbohydrate-containing plant raw materials and wastes based on the use of a combined (hybrid) method

Just like the rate of any chemical transformation, the rate of enzymatic reactions depends on many factors. The main ones are the following - the time of the enzymatic reaction (incubation time), the temperature and pH of the environment. As the incubation time of the enzyme increases, the rate of the reaction decreases. This can occur for a variety of reasons, the main ones being a decrease in substrate concentration, an increase in the rate of the reverse reaction (due to product accumulation), inhibition of the enzyme by the reaction product, and denaturation of the enzyme. In kinetic studies, measurements of the initial reaction rate are performed. According to the Want-Goff rule for chemical reactions, a 10°C increase in temperature leads to a 2-4 fold increase in reaction rate. The rate of enzymatic reaction increases with increasing temperature, reaching a maximum at some optimal temperature (temperature optimum of the enzyme), and then falls to zero. It should be noted that the temperature coefficient of rate increase for enzymatic reactions is smaller than for ordinary chemical reactions: the rate increases no more than twofold for every 10°C increase in temperature. The decrease in reaction rate to zero is caused by inactivation of the enzyme due to denaturation of its protein part. Most enzymes are inactivated at temperatures of 40-50°C. Some enzymes are inactivated at temperatures close to 0°C. Enzymes that are glycoproteins are characterized by high thermostability, since the carbohydrate fragment gives the protein thermostability. Such enzymes work in microorganisms living in hot springs.

A change in the pH of the medium leads to a change in the degree of ionization of the acidic and basic groups of both the enzyme active center and the substrate itself. Consequently, a change in pH affects the affinity of the substrate for the enzyme active center and the catalytic mechanism of the reaction. Usually, the dependence of the rate of enzymatic reaction on the pH of the medium is bell-shaped, because for each enzyme there is an optimal pH value at which the enzyme exhibits the greatest catalytic activity (enzyme pH optimum). The optimum pH value corresponds to the best substrate binding by the enzyme and the highest rate of catalysis.

For the planned experiments on enzymatic hydrolytic hydrolysis and hydrogenation were subjected to beer pellets of "Shymkentpivo" LLP.

The catalysts used were BruzheimBSX, The enzymatic activities of Bruzheim BGX (producer of *Trichodermalongibrachiatum*) are summarized in Table 5.

Table 5 - Characterization of the enzyme preparation

Name	Processing parameters	Activity

drug	temperature, T	pH	
BruzeimBCX (mushroom hemicellulase)	30-50	3,5-5,5	Xylanase: 6500+5 % units of KS/cm3; β-Glucanase: 1450+5 % units of SG-HcS/cm3; Cellulase: 1500 units +5 % units of CMC/cm3;

The enzymatic hydrolytic hydrolysis and hydrogenation of each sample was carried out in aqueous medium (active acidity was adjusted with phosphoric acid and was in the range of 4.8-4.9 pH units). The substrate concentration in all experiments was 45.0 g/L. The composition of enzyme preparations "BruzaimVSH" was used as catalysts, applied in the amount of 0.03 g of enzyme per 1 g of substrate in the

the beginning of fermentation. To carry out the hydrogenation-hydrolysis process, accurately weighed on analytical scales amounts of beer grist, enzyme, phosphoric acid and the required amount of water were placed in the fermenter. After a prescribed time interval, the process was stopped and analyzed for sugar alcohol content and cellulose conversion by liquid chromatography.

Dependences of the rate of enzymatic hydrolytic hydrolysis and hydrogenation of beer pellets on the time of the process are shown in Figure 9 and Table 6.

Table 6 - Dependence of the rate of enzymatic hydrolytic hydrolysis and hydrogenation of beer pellets on the process time

№/№	τ, mi н	Conversion rate,%	Xylitol selectivity,%	Mannitol selectivity,%	Total yield,%	Xylose selectivity
1.	80	34,0	1,9	1,0	23,4	19,1
2.	160	38,1	2,9	1,4	34,5	21,8
3.	240	55,0	3,6	1,8	47,1	35,0
4.	320	49,7	3,0	1,6	40,9	33,3
5.	400	47,6	2,3	1,3	37,8	31,1

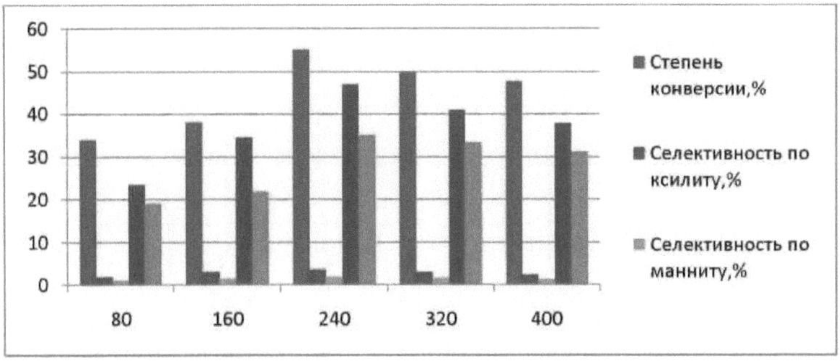

Figure 9 -Hydrogenation of beer pellets as a function of process time

Table 7 shows the experimental data on the study of the regularities of change in the rate of enzymatic hydrolytic hydrolysis and hydrogenation of beer yeast from the reaction time. The reaction time was varied from 80 to 400 minutes. The optimum reaction time for the catalytic conversion of beer pellets under our selected conditions was determined to be 240 minutes. Up to this minute of reaction time, the conversion of beer pellets increases smoothly, and after this time its values are within the margin of error. The same pattern is observed with the selectivity index for xylitol. However, the selectivity for xylitol and mannitol is much lower than that of chemical hydrolytic hydrolysis and hydrogenation. This is due to the predominance of the hydrolysis reaction rate over the hydrogenation reaction rate. This is evidenced by the large values of selectivity for xylose (from 19.1 to 35.0%). To increase the rate of hydrogenation reaction, we additionally added 0.03 g of microscopic fungal strain of Trichoderma genus to the fermenter. The results of the experiments are summarized in Table 7 and Figure 10.

Table 7 - Dependence of the rate of enzymatic hydrolytic hydrolysis and hydrogenation of beer pellets on the time of the process in the presence of complex enzymes.

№/№	τ, mi н	Conversion rate, %	Xylitol selectivity,%	Mannitol selectivity,%	Total yield,%	Xylose selectivity
1.	80	33,4	14,9	1,2	21,4	1,9

2.	160	35,8	15,9	1,0	31,3	2,8
3.	240	42,7	17,8	1,1	37,1	3,0
4.	320	39,2	14,7	1,3	35,7	2,9
5.	400	47,6	2,3	1,3	37,8	31,1

The data of Table 7 show that the selectivity for xylitol increased from 3.6 to 17.8%, while the selectivity for xylose decreased from 35.0 to 3.0%. This pattern indicates the predominance of the hydrogenation reaction rate over the hydrolysis reaction rate. I.e. we realized the process of combined (hybrid) enzymatic process of hydrolytic hydrolysis and hydrogenation under these conditions.

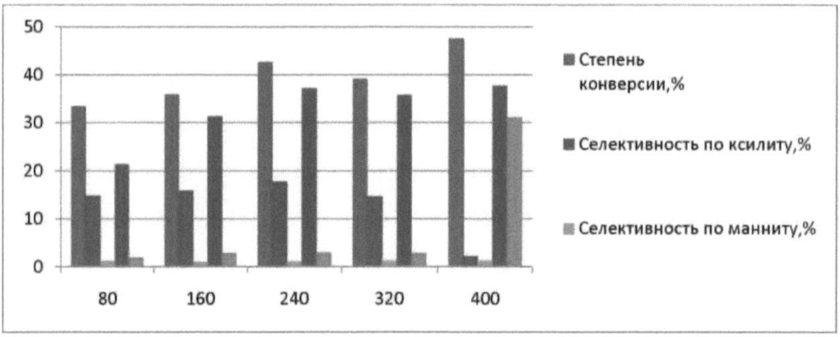

Figure 10

In studying the effect of temperature of the enzymatic hydrolytic hydrolysis and hydrogenation process in the presence of complex enzymes on the conversion of beer pellets and selectivity for xylitol and mannitol, it was shown that as the temperature increased from 30 to 50° C, the conversion of beer pellets increased from 12.7 to 42.7% (Table 8,9 and Fig. 11.). The selectivity for xylitol increased from 7.7 to 17.8% with increasing temperature(30-50° C) and decreased to 7.0% when the temperature was increased to 70° C. The decrease in xylitol selectivity is due to the fact that at temperatures above 70° C the inactivation process of the enzymes used occurs. Table 4 shows that the optimum temperature of the experiment is 50° C, because at this temperature we obtained the maximum selectivity for xylitol and mannitol.

Table 8 - Effect of experiment temperature on the process of enzymatic hydrolytic hydrolysis and hydrogenation of beer pellets

№/№	T C◦	Ogepen	Selectivity	Selectivity	Total
		conversions,%	xylitol,%	mannitol,%	yield,%
1.	30	12,7	7,7	1,8	10,2
2.	40	28,3	14,3	2,0	16,1
3.	50	42,7	17,8	1,1	18,9
4.	60	34,3	12,1	1,1	14,8
5.	70	19,7	7,0	1,2	9,0

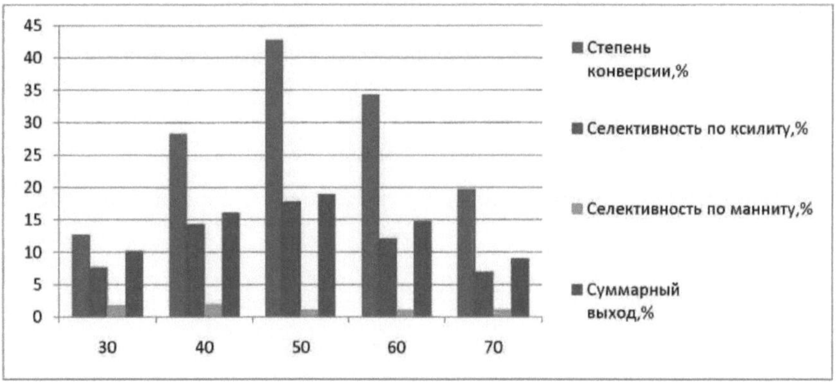

Figure 11

Table 9 - Effect of pH on the process of enzymatic hydrolytic hydrolysis and hydrogenation of beer pellets

№/№	pH	Conversion rate,%	Xylitol selectivity,%	Mannitol selectivity,%	Total yield,%
1.	4,0	15,2	7,5	1,2	9,3
2.	4,5	30,0	11,3	1,2	12,5
3.	5,0	42,7	17,8	1,1	18,9
4.	5,5	35,8	10,5	2,9	11,7

5.	6,0	19,1	5.5	2,4	8,1

During the study of the influence of pH on the process of enzymatic hydrolytic hydrolysis and hydrogenation of beer pellets on conversion and selectivity for polyols it was found that the highest values of selectivity for xylitol - 17.8% and conversion - 42.7% are observed in the case of using pH value equal to 5.0. Changing the pH of the medium leads to a change in the degree of ionization of acidic and basic groups of both the active center of the enzyme and the substrate itself. Consequently, a change in pH affects the affinity of the substrate for the active center of the enzyme and the catalytic mechanism of the reaction. The dependence of the rate of enzymatic reaction on the pH of the medium has the form of an extremum, because for each enzyme there is an optimal pH value at which the enzyme exhibits the greatest catalytic activity (enzyme pH optimum). The pH value in the optimum corresponds to the best binding of substrate by the enzyme and the highest rate of catalysis. In our case, this value is 5.0.

Thus, we have determined the optimal conditions of the process of enzymatic hydrolytic hydrolysis and hydrogenation of beer pellets: pH - 5.0, temperature 50° C, duration of the reaction - 3 hours

3.3 Development of technology for synthesis of a new generation of hydrolytic and liquid-phase hydrogenation catalysts and selection of optimal process parameters in their presence

3.3.1Technology of synthesis of a new generation of catalysts for hydrolytic and liquid-phase hydrogenation

The technological scheme of production and testing of alloy skeletal catalysts developed by us consists of: a high-frequency melting furnace, a batch reactor and a column-type flow unit. The principal technological scheme of catalyst production is shown in Fig. 12.

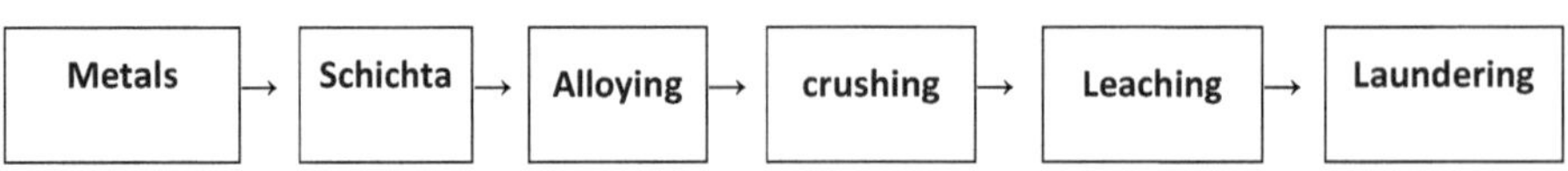

Figure 12 - Basic technological scheme of catalyst production

The alloys were prepared in a high-frequency melting furnace of the brand according to the technology developed by us. A calculated amount of Al in the form of ingots was placed in a quartz crucible and gradually heated to 1000 - 1IOO^0 C, then a calculated amount of nickel, copper or cobalt and ferroalloy additives in the form of chips or powder were introduced. As a result of the exothermic reaction, the temperature of the melt was raised to 1700- 1800^0 C, which was stirred for 3-5 minutes by an induction field. In graphite liners, the alloy was air cooled and milled to 0.25 mm grains. The alloys were activated by leaching with 1050% aqueous solution of caustic sodium taken in the volume of 40 cm^3 per 1 g of alloy on a boiling water bath for 1 hour, after which the catalyst was washed from the alkali with water to neutral reaction according to phenolphthalein.

In order to develop the technology for synthesis of new generation catalysts, we obtained the following catalytic systems: Ni-Si-FS,FS- Al,FS--Si

3.3.2Selection of optimal process parameters for hydrolytic and liquid-phase hydrogenation in the presence of developed catalysts

3.3.2.1 Selection of optimal process parameters for chemical hydrolytic hydrolysis and hydrogenation of beer pellets

The study of the influence of various technological parameters on the process is of great theoretical and practical importance, since it allows not only to present a picture of the interaction of reacting molecules, but also, ultimately, to choose the most optimal catalyst composition and conditions for the process under industrial conditions. At the same time, the studies make it possible to make assumptions about the reaction mechanism.

The study of chemical hydrolytic hydrolysis and hydrogenation process of beer pellet was investigated:

- The effect of experiment temperature;
- influence of the nature of the promoting ferroalloy
- the effect of the nature of the reducing agent

- effect of the nature of the acid used
- influence of the nature of the base metal of the cellulose catalyst. The product yield was determined as the ratio of the mass of the formed product to the initial mass of cellulose.

When studying the effect of process temperature on beer pellet conversion and selectivity for xylitol and mannitol, it was shown that with increasing temperature from 75 to 175° C, beer pellet conversion increased from 20.5 to 64.8% (Table 9,11,12 and Fig. 13,14,15) This is due to the fact that at high temperatures the concentration of H_3 O and OH ions in water increases, which causes an increase in the acidity of the medium required for hydrolysis of beer pellet. The selectivity for xylitol increased with increasing temperature(75-125° C) from 12.3 to 19.8% and decreased to 10.3% with increasing temperature from 150 to 175° C. Decrease of selectivity for xylitol is caused by the fact that at temperatures above 125° C the rates of sugars decomposition processes (hydrogenolysis) with formation of lower polyols begin to increase. Table 9 shows that the optimum temperature of the experiment is 125° C, because at this temperature we obtained the maximum selectivity for xylitol and total yields of sorbitol and mannitol.

Table 9 - Effect of experiment temperature on the process of chemical hydrolytic hydrolysis and hydrogenation of beer pellets.

№/№	T C·	Ogepen	Selectivity	Selectivity	Total
		conversions,%	xylitol,%	mannitol,%	yield,%
1.	75	20,5	12,3	2,4	13,6
2.	100	48,4	15,5	2,3	17,4
3.	125	53,0	19,8	2,1	21,8
4.	150	62,2	14,0	1,1	14,8
5.	175	64,8	10,3	1,6	11,0

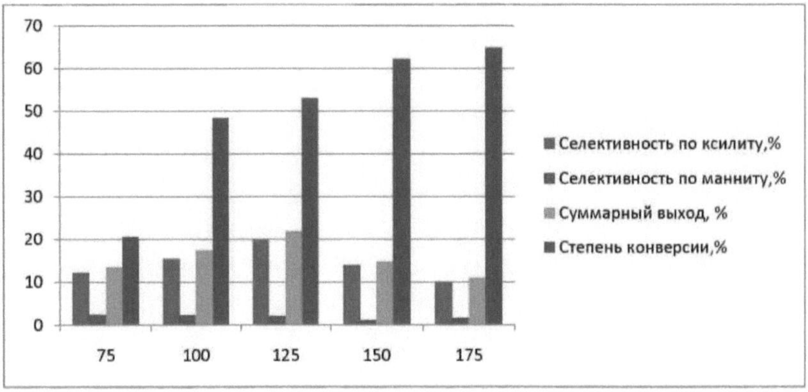

Figure 13 - Experimental conditions: Ni-FS catheter, 24% isopropanol, 120 minutes

During the study of the influence of the nature of the acid used on the process of chemical hydrolytic hydrolysis and hydrogenation of beer pellets on the conversion and selectivity for polyols it was found that with the highest values of selectivity for xylitol - 19.8% and conversion - 75% are observed in the case of using sulfuric acid. When phosphoric acid is used, the lowest values of conversion -55.7% and selectivity for xylitol - 12.3% are observed. According to the activity in the process of chemical hydrolytic hydrolysis and hydrogenation of beer grit, the studied acids form the following series:

$$H_2SO_4>HNO_3>H_2SO_3>H_3PO_4$$

Table 10 - Effect of acid nature on the process of chemical hydrolytic hydrolysis and hydrogenation of beer pellets.

№/№	acid	Conversion rate,%	Xylitol selectivity,%	Mannitol selectivity,%	Total yield,%
1.	HNO3	67,4	16,4	1,7	17,5
2.	H2SO4	75,0	19,8	2,1	21,8
3.	H2SO3	65,2	17,8	3,0	22,6
4.	HCI	70,6	17,3	2,9	18,5
5.	H3PO4	55,7	12.3	2,4	15,0

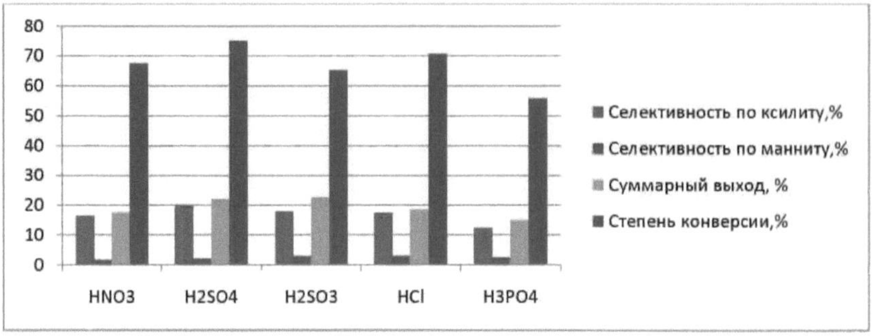

Figure 14- Experiment conditions: cat Ni-FS, 24% isopropanol, 120 minutes, Ton=125° C.

Table 11 - Effect of the nature of reducing agent on the process of chemical hydrolytic hydrolysis and hydrogenation of beer pellets

№/№	acid	Conversion rate,%	Xylitol selectivity,%	Mannitol selectivity,%	Total yield,%
1.	methanol	0	0	0	0
2.	ethanol	1,2	0	0	0
3.	Isopro-panol	75,0	19,8	2,1	21,8
4.	butanol	3,6	0,3	0,9	1,5
5.	glycerine	5,7	2.3	2,4	5,0

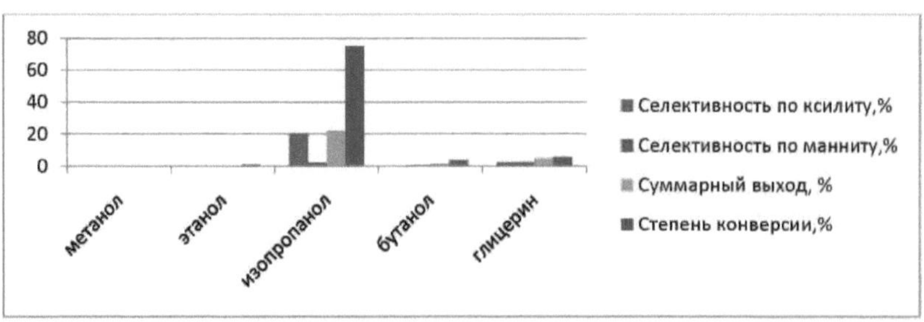

Figure 15- Experiment conditions: Ni-FS catheter, H_2SO_4, 120 minutes, Ton= 125 C°

The second part of the process of chemical hydrolytic hydrolysis and hydrogenation of beer pellets, namely the hydrogenation of the formed xylose to xylitol or mannitol, depends on the efficiency of the reducing agent. From Table 12, which summarizes the results of the study, it can be seen that the beer pellet conversion and xylitol selectivity have the highest values when isopropanol is used. When ethanol and methanol are used, the process of chemical hydrolytic hydrolysis and hydrogenation is practically absent. In case butanol and glycerol are used, the reaction rate is negligible and the process cannot be used for realization (Fig. 16).

Table 12 - Dependence of the rate of the process of chemical hydrolytic hydrolysis and hydrogenation of beer pellets on the content of ferroalloy.

№/№	Contents. ferroalloy,%	Conversion rate,%	Selective immunity xylitol,%	Selective immunity mannitol,%	Total yield,%
1.	1,0	38,7	12,9	2,3	18,7
2.	3,0	50,0	16,5	2,3	20,0
3.	5,0	68,5	18,0	2,0	20,0
4.	7,0	75,0	19,8	2,1	21,8
5.	10,0	73,3	18,7	2,3	20,9

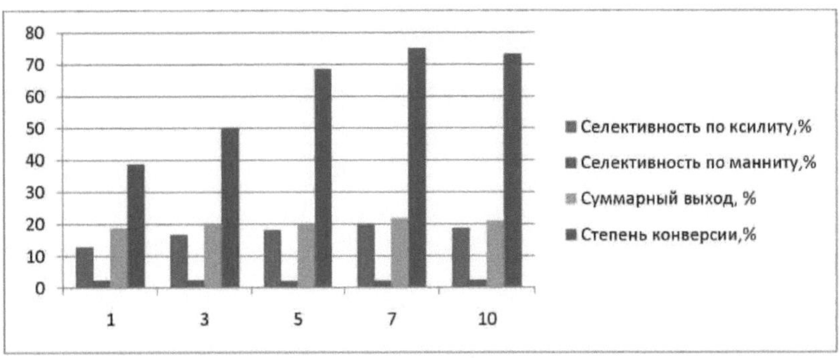

Figure 16 - Experiment conditions: Ton= 125° C, , 120 minutes

As can be seen from Table 12, in the range of ferroalloy content in the promoted nickel catalyst from 1.0 to 7.0%, an increase in conversion rate from 38.7 to 75% and in xylite selectivity from 12.9 to 19.8% was observed.When the ferroalloy content in the promoted nickel catalyst increased above 7.0 wt.%, a decrease in xylite selectivity was observed with a simultaneous increase in the yield of pentodes, tetrodes and lower polyols.

Ferroalloy-promoted catalysts of three types: nickel, copper and cobalt were used for realization of the process of chemical hydrolytic hydrolysis and hydrogenation of beer pellets.These catalysts in the studied process showed different activity, which was evaluated by the degree of conversion and selectivity for polyols (Table 13).In the course of the study it was shown that such a difference is due to structural and physicochemical properties of these catalysts.For example, as shown in the phase and particle size distribution, porous

Table 13 - Dependence of the rate of the process of chemical hydrolytic hydrolysis and hydrogenation of beer pellets on the nature of the ferroalloy catalyst

№№	nature. metal%	Conversion rate,%	Xylitol selectivity,%	Mannitol selectivity,%	Total yield,%
1.	Ni(FS)	75,0	19,8	2,1	21,8
2.	Cu(FS)	43,1	8,8	2,0	11,3
3.	Co(FS)	54,0	12,2	2,1	14,5
4.	/Al_2O_3(FS)	91,4	21,8	3,4	22,0
5.	/Al_2O_3(FS)	83,5	20,6	3,0	21,3

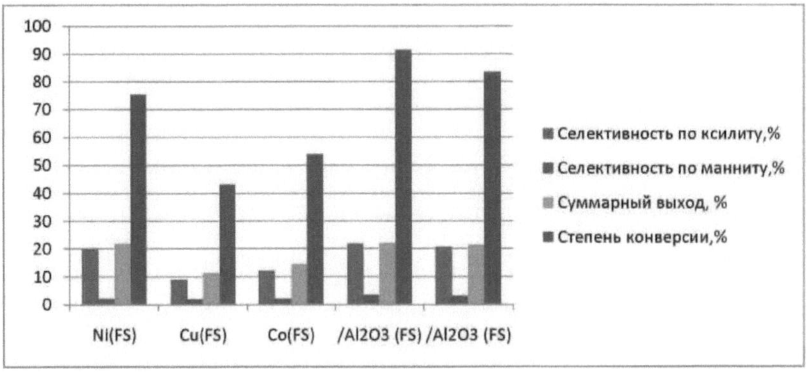

Figure 17.

3.3.2.2 Continuous hydrolytic hydrogenation of beer yeast polysaccharides in a batch column unit

Some alloy copper, cobalt and nickel catalysts promoted by ferroalloys were tested in the process of hydrogenation of polysaccharides of beer pellets in order to develop optimal catalytic systems for xylitol production at the column batch plant. As a raw material in the batch column unit was used beer pellet of "Shymkentpivo" LLP. The results of the studies are given in Table 14.

Table 14 - Results of simultaneous hydrolysis and hydrogenation of beer pellets in a batch column plant. Conditions: pressure - 4MPa, temperature = 100°C, catalyst granule size 6-8mm, pH=3.5

catalyst	Xylitol yield, %
Cu - Al	69,5
Cu- Al - FS	80,3
Co - Al - FS	87,5
Ni- Al- FS	92,8

Table 14 shows that the investigated catalysts exhibit absolute activity on xylite, in terms of activity they form the following series: Cu- Al - FS< Co - Al -FS< Ni- Al-FS

One of the main process parameters is hydrogen pressure, the increase of which increases the hydrogen concentration in the solution and on the catalyst surface. Table 15 shows the results of the study of the effect of hydrogen pressure on xylitol yield in the combined (hybrid) process of xylitol production from beer pellets.

Table 15 shows that with increasing hydrogen pressure in the system, the rate of its reproduction on the catalyst surface increases and this contributes to an increase in the reaction rate and xylite yield. The xylite yield at 6 MPa for the most active Ni-Al-FS catalyst is 97.0%. The optimum hydrogen pressure for the process of continuous hydrolytic hydrogenation of polysaccharides of beer mash in a batch column unit is 5 MPa, and increasing the pressure up to 6 MPa has no appreciable effect.

Table 15 - Influence of hydrogen pressure on xylite yield during simultaneous hydrolysis and hydrogenation of beer pellets in a batch column plant. Conditions: temperature =120^0 C, catalyst granule size: 6-8mm, pH=3.5

catalyst	PH2, MPa	Xylitol yield, %
Cu-Al	2	67,3
	4	70,5
	6	73,1
Cu-Al-FS	2	81,5
	4	84,3
	6	85,4
Co-Al- FS	2	90,7
	4	93,4
	6	95,1
Ni-Al-FS	2	91,6
	4	95,5
	6	97,0

Table 16 summarizes the results of the study of the effect of the experimental temperature on the yield of xylitol. The experimental data show that the highest xylose yields are achieved at 80°C.

Table 16 - Effect of temperature on xylitol yield from simultaneous hydrolysis and hydrogenation of beer pellets in a batch column plant

catalyst	T,°C	Xylitol yield, %
Cu- Al	60	60,4
	80	71,0
	100	71,3
Cu- Al- FS	60	75,8
	80	88,8
	100	85,9
Co-Al-FS	60	84,8
	80	95,9
	100	94,4
Ni- Al-FS	60	86,5
	80	97,4
	100	93,8

Optimal catalysts should be characterized not only by high activity and selectivity, but also by stability, which depends both on the mobility of active hydrogen (reproduction rate) and on the mechanical strength of the initial alloy and catalyst. It is known that catalysts with promoter metals are more than twice as active and stable as catalysts without promoter.

Alloy nickel catalysts obtained from multicomponent alloys have been investigated. Ferroalloys ferromolybdenum (FMo), ferrotitanium (ΦTi) and ferrosilicocalcium (FSC) were used as additives.

Table 17 shows that the investigated nickel catalysts show high activity on xylite, the rate of formation of the latter increases differently depending on the chemical nature and amount of modifying ferroalloys. The maximum activity is shown by the catalyst from the alloy with 5 wt. % FMo. The yield of xylite on it IOO^0 C and 10 MPa at 60 minutes is 94.4%, which is more than an order of magnitude higher than on skeletal nickel without additives.

Table 17 - Effect of the amount of introduced ferroalloy additive on the activity of multicomponent nickel catalyst PH_2 = 6MPa

Additive, %	t , $C_{оп}^0$	Xylitol yield (%) in time (min)				W- **104** mol gkt/min
		10	20	40	60	
1	2	3	4	5	6	7
ΦTiNi - Al - ΦTi						
1,0	50	19,5	26,1	33,5	45,8	14,4
	100	23,4	32,7	45,7	60,6	19,3
3,0	50	21,0	27,8	36,6	49,8	15,3
	100	28,8	39,6	58,6	72,7	22,6
5,0	50	22,2	29,9	39,1	53,4	16,2
	100	31,6	43,0	61,8	79,6	24,2
7,0	50	23,6	30,9	40,7	58,1	16,8
	100	34,6	43,5	63,5	81,4	25,1
10,0	50	23,2	31,5	39,0	59,8	17,2
	100	38,7	45,9	64,8	83,6	25,9
ΦMoNi-Al - ΦMo						
1,0	50	20,1	29,3	39,9	49,0	16,8
	100	30,3	43,9	57,8	72,8	27,5
3,0	50	24,3	33,1	41,4	54,6	20,4
	100	33,5	61,4	64,5	84,8	32,1
5,0	50	30,4	39,0	49,5	67,1	22,6
	100	35,7	66,9	69,7	91,4	33,6
7,0	50	29,8	40,2	51,4	70,7	23,4
	100	40,2	56,2	71,4	92,8	34,2
10,0	50	29,8	40,7	52,6	71,4	23,7

	100	40,5	56,9	72,6	94,4	34,7
FGC Ni - Al - FGC						
1,0	50	10,3	13,6	21,3	27,0	8,1
	100	24,3	30,5	42,6	55,3	14,6
3,0	50	14,2	17,7	27,1	36,1	10,8
	100	26,3	33,6	47,4	63,6	16,1
5,0	50	18,5	23,8	33,5	43,0	12,0
	100	28,7	36,4	51,8	66,2	17,1
7,0	50	11,7	17,0	25,6	35,4	10,4
	100	25,6	32,9	46,4	59,4	15,8
10,0	50	11,5	15,6	23,2	30,6	9,4

3.3.2.3 Selection of optimal process parameters for chemical hydrolytic hydrolysis and hydrogenation of guza-pai cellulose

We have shown earlier (15) the possibility of obtaining cellulose from guza-paya by autohydrolysis. We used this cellulose to realize a combined (hybrid) hydrolysis-hydrogenation process to produce sorbitol.

In the process of chemical hydrolytic hydrogenation of guza-pai cellulose into sorbitol, the influence of the experiment temperature was studied in the range of 140-220° C. Table 18 shows that the optimum experiment temperature is 180° C, because at this temperature we obtained the maximum sorbitol selectivity and total yields of sorbitol and mannitol (Fig. 18).

At temperatures 140 and 160° C, the cellulose conversion (20.524.4%), sorbitol selectivity (11.4-14.6%) and total yield (14.916.1%) are much lower than those at 180° C. Although at temperatures of 200-220° C the conversion of guza-pai cellulose increases significantly (74.476.2%), a decrease in sorbitol selectivity (9.1-9.8) and total yield of 10.0-10.8% is observed. This is explained by the appearance of other substances in the solution, such as polyols with the number of atoms below five.

Table 18 - Effect of experiment temperature on the process of chemical hydrolytic hydrogenation of guza-pai cellulose

№/№	T C◦	('tepen conversions,%	Sorbitol selectivity,%	Mannitol selectivity,%	Total yield,%
1.	140	20,5	11,4	2,4	14,9
2.	160	24,4	14,6	2,9	16,1
3.	180	53,0	21,8	3,0	22,6
4.	200	76,2	9,8	1,3	10,8
5.	220	74,4	9,1	1,2	10,0

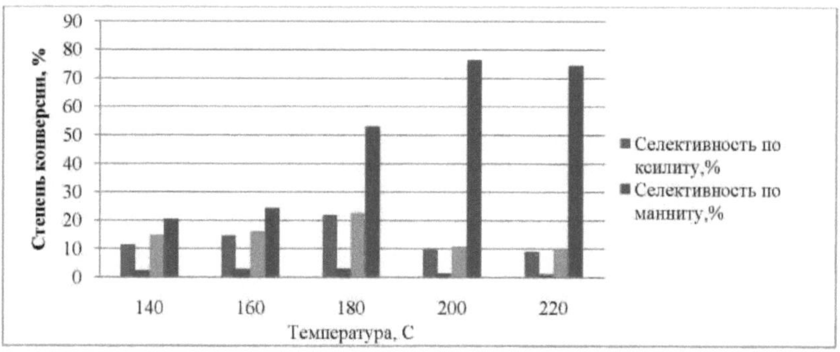

Figure 18 - Experiment conditions: 0.5 g. 3% Ni/Al O_{23} (ΦS), 60 min, P_{H_2} =6.0 MPa,

Table 19 - Effect of hydrogen pressure on the process of chemical hydrolytic hydrogenation of guza-pai cellulose

№/№	R 12, MPa	Conversion rate,%	Sorbitol selectivity,%	Mannitol selectivity,%	Total yield,%
1.	2,0	42,3	13,0	1,7	13,5
2.	4,0	51,6	13,5	1,8	14,2
3.	6,0	53,0	21,8	3,0	22,6
4.	8,0	76,6	18,9	2,8	31,3
5.	10,0	77,8	16.5	2,7	32,8

The influence of hydrogen pressure on the process of chemical hydrolytic hydrogenation of cellulose guza-pai was investigated in the range from 2.0 to 10.0 MPa. Table 19 shows that with increasing hydrogen pressure, the degree of cellulose conversion increases from 42.3 to 77.8 %. However, sorbitol selectivity has a maximum at a pressure of 6.0 MPa. That is, the share of the desired product - sorbitol with increasing hydrogen pressure above 6.0 MPa decreases due to the formation of five-atom alcohols. This is expressed in the growth of the total yield of polyols. Thus, we chose 6.0 MPa as the optimal pressure (Fig.19, 20).

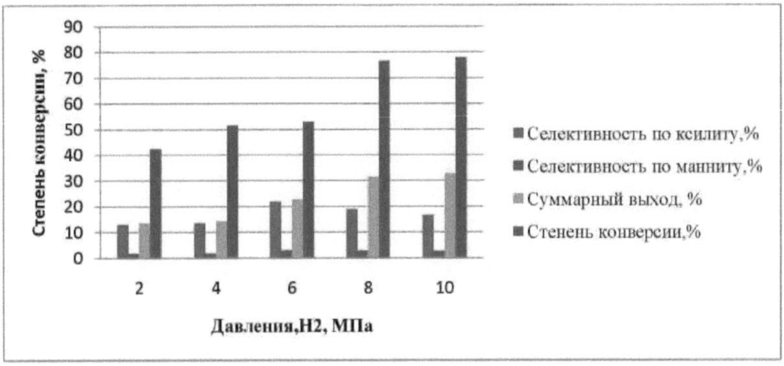

Figure 19 - Experiment conditions: 0.5 g. 3% Ni/Al$_2$o$_3$(ФS), 60 min, Top= 180 C°

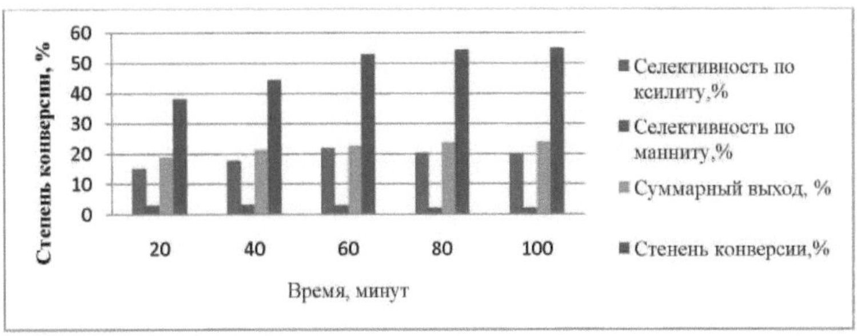

Figure 20

As can be seen from Table 20, when the nickel content in the catalyst increased above 3.0 wt.%, a decrease in the yield of sorbitol was observed with a simultaneous increase in the yield of pentodes, tetrodes and lower polyols. The conversion rate decreased by 9% at 5% and on average by 21% at 7-10% nickel content. Reducing the nickel percentage below 1% is undesirable because it leads to instability of the obtained experimental data (Figs. 21,22 and Table 21).

Table 20 - Dependence of the rate of the process of chemical hydrolytic hydrogenation of celluloseguza-pai on the composition of the applied catalyst

№/№	Nickel content,%	Conversion rate,%	Sorbitol selectivity,%	Mannitol selectivity,%	Total yield,%
1.	1,0	53,0	24,3	3,3	25,2
2.	3,0	53,0	24,3	3,3	25,2
3.	5,0	48,1	19,3	3,0	23,3
4.	7,0	42,0	18,7	3,	22,0
5.	10,0	40,5	17,5	3,0	21,3

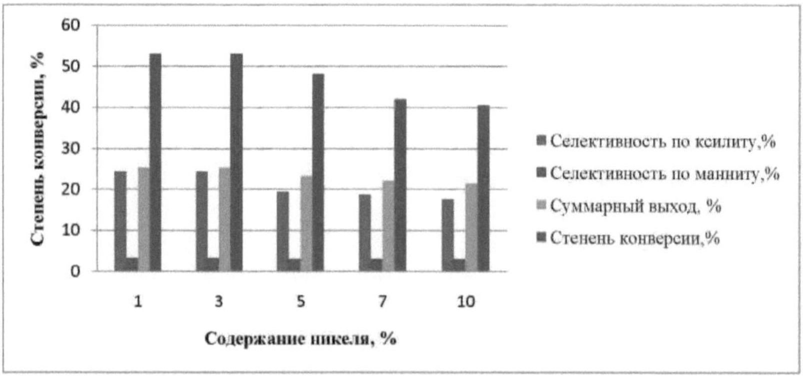

Figure 21 - Experiment conditions: Top = 180° C, Rn_2 =6,0 MPa, 60 minutes

Table 21 - Dependence of the rate of the process of chemical hydrolytic hydrogenation of celluloseguza-pai on the nature of the applied catalyst

№№	Nature. Metal%	Conversion rate,%	Sorbitol selectivity,%	Mannitol selectivity,%	Total yield,%
1.	Ni/Al_2O_3(FS)	70,0	24,3	3,3	25,2

2.	Cu/Al2O3(FS)	70,0	24,3	3,3	25,2
3.	CoMJ2O3(FS)	81,3	23,0	3,1	23,3
4.	/Al2O3(FS)	91,4	21,8	3,4	22,0
5.	/Al2O3(FS)	83,5	20,6	3,0	21,3

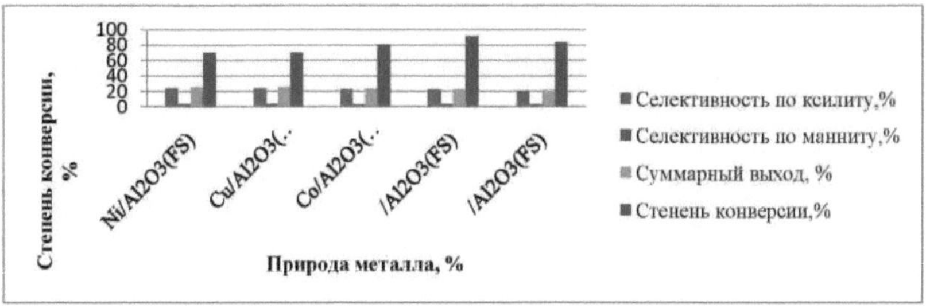

Figure 22

3.4 Study of kinetics of combined processes of hydrolytic hydrolysis and hydrogenation of guza-pai cellulose

The process was carried out at hydrogen pressure of 4 -6 MPa in 0.01 - 1% solution of $H_3 PO_4$ at temperature 60 - 100° C, process duration from 30 to 120 min in the presence of suspended ferroalloy-promoted granular nickel catalyst.

As follows from the data of Figures 23 and 24, the dependence of the degree of conversion of guza-pai cellulose on the concentration of $H_3 PO_4$ has an extreme character, the highest degree of conversion is observed at the acid concentration equal to 0.8%.

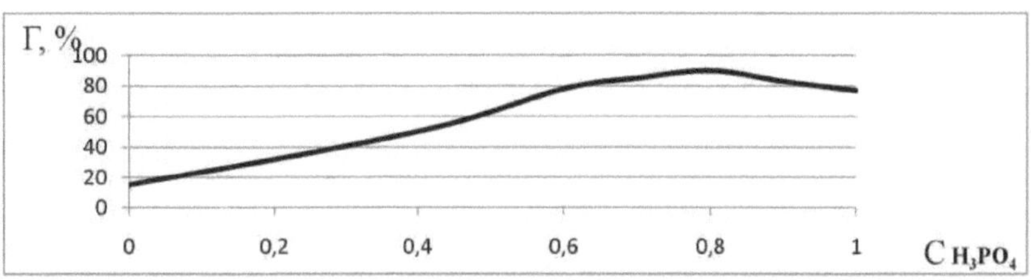

Figure 23 - Kinetic curves of hydrolysis of cellulose guza-pai at 100^0 C and Rn $4MPa_2$

Smaller concentrations of acid are not enough to initiate autohydrolysis, and large to use economically and technologically inexpedient, because in this case there is a simple acid hydrolysis, which consumes phosphoric acid, and its excess prevents 85

the free formation of sorbitol.

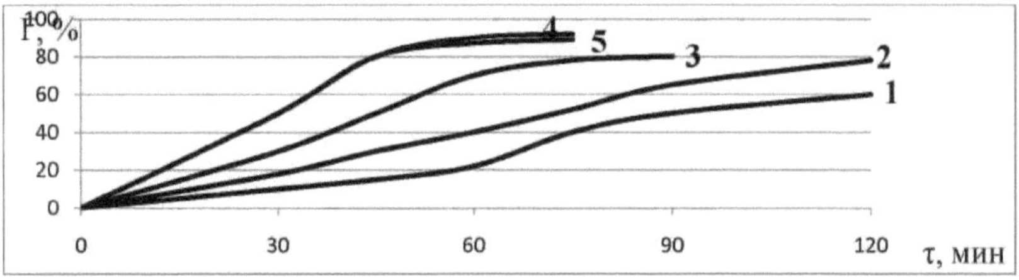

Curve designation, temperature,^{0} C : 1-20; 2-40; 3-60; 4-80; 5-100.

Figure 24 - Kinetic curves of hydrolysis of guza-pai cellulose at P_{H_2} 4MPa in 0.8% H_3PO_4 .

In conditions of joint hydrolytic hydrolysis and hydrogenation at increased pressures it is possible to reduce the temperature of the process from 100 to 80^0 C, because in this case not only by 2% increases the yield of sorbitol, but also to save thermal energy.

Thus, the principal possibility of developing a new technology for the production of sorbitol from guza-pai cellulose by combining the processes of hydrolysis and hydrogenation in one apparatus has been shown.

The following catalysts were studied for the synthesis of new generation catalysts for the process of chemical hydrolytic hydrolysis and hydrogenation of beer pellets: ferroalloy-promoted nickel-silicon, ferroalloy-aluminum and ferroalloy-silicon catalysts. These catalysts showed different activity in the studied process, which was evaluated by the degree of conversion and selectivity for polyols (Table 22).During the study, it was shown that this difference is due to the structural and physicochemical properties of these catalysts. For example, phase and particle size distribution, porosity, surface state and adsorption properties. It was revealed that the highest selectivity for xylitol (19.8) and degree of conversion (75%) is characterized by nickel catalyst promoted by ferroalloys.The lowest selectivity for xylitol (8.8%) and the degree of conversion (43.1%) was characterized by copper catalyst promoted by ferroalloys.In terms of activity these contacts form the following series: Ni-FS>Co-FS>Cu-FS.

Table 22 - Dependence of the rate of the process of chemical hydrolytic hydrolysis and hydrogenation of beer pellets on the nature of ferroalloy catalyst

№№	nature. Metal%	Conversion rate,%	Xylitol selectivity,%	Mannitol selectivity,%	Total yield,%
1.	Ni(FS)	75,0	19,8	2,1	21,8
2.	Cu(FS)	43,1	8,8	2,0	11,3
3.	Co(FS)	54,0	12,2	2,1	14,5
4.	$/Al_2O_3$(FS)	91,4	21,8	3,4	22,0
5.	$/Al_2O_3$(FS)	83,5	20,6	3,0	21,3

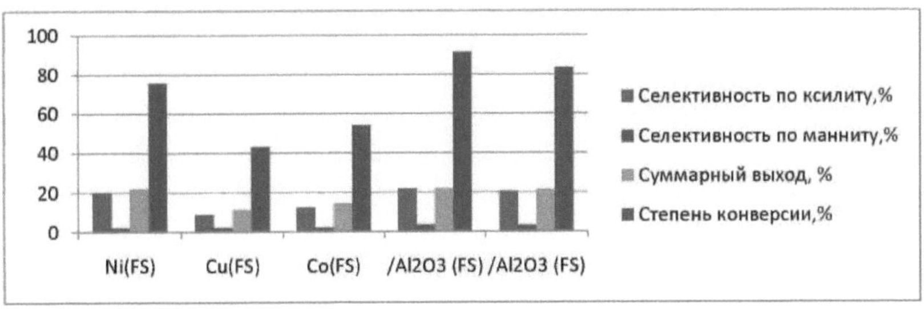

Figure 25

In terms of addition to the previously developed new generation catalysts and tested in the process of chemical and enzymatic hydrolytic hydrolysis and hydrogenation of beer pellets and guza-paya (described in Chapter 3), the following catalysts were additionally developed and investigated: ferroalloy-promoted nickel-aluminum nickel-silicon, aluminum, silicon, and nickel on aluminum oxides. These catalysts in the studied process showed different activity, which was evaluated by the degree of conversion and selectivity for polyols (Table 23).The table shows the results of the study of catalytic activity of the developed catalysts in the process of chemical hydrolytic hydrolysis and hydrogenation of beer pelletsIn the course of the study it was shown that such a difference is due to the structural and physicochemical properties of these catalysts. For example, phase and particle size distribution, porosity, surface state and adsorption properties.It was found that the highest selectivity for xylitol (22.2%) and conversion rate (77.4%) is characterized by ferroalloy-promoted nickel-silicon catalyst.The lowest selectivity for xylitol (19.8%) and conversion rate (52.4%) is ferroalloy-promoted silicon catalyst. In terms of activity, these contacts form the following series: Ni- Si-FS>Ni(FS)>FS-AL>Ni/Al_2O_3(FS) >.FS-Si.

Table. 23 .Results of the catalytic activity of the developed catalysts in the process of chemical hydrolytic hydrolysis and hydrogenation of beer pellets on the developed catalysts

№№	nature. metal%	Conversion rate,%	Xylitol selectivity,%	Mannitol selectivity,%	Total yield,%	
1.	Ni(FS)	75,0	19,8	2,1	21,8	
2.	Ni-Si-FS	77,4	22,2	2,0	24,4	
3.	FS-AL	65,3	20,3	2,0	23,7	
4.	FS-Si	52,4	21,8	3,4	22,0	
5.	≡Al_2O_3(FS)	63,5	20,6	3,0	21,3	

The further goal of our research will be to determine the physicochemical properties of contacts and identify the correlation between catalytic, adsorption and physicochemical properties, as well as to study the kinetics and mechanism of processes, quantum-chemical calculations and modeling, which will allow us to develop original technologies for the production of xylitol and sorbitol, taking into account the requirements of the technical specification .

CONCLUSION

1 . Development of a one-stage combined (hybrid) process for conversion of carbohydrate-containing raw materials and wastes (guza-pai pulp and beer pellets) into sorbitol and xylitol has been carried out

2 The technology of the process of hydrolysis-hydrogenation of guza-pai cellulose into sorbitol has been developed. The optimum parameters of the process were determined.

3 . For the first time the technology of depolymerization of beer pellets to xylitol was substantiated and experimentally confirmed by establishing the optimal modes of the process apparatus.

4 . The kinetic characteristics of the enzymatic combined (hybrid) process of sorbitol production from beer pellets have been studied. The conditions of synthesis of the reaction product close to theoretically possible have been established.

5 The developed technologies of chemical and enzymatic combined (hybrid) hydrolysis-hydrogenation of guza-paya cellulose and beer pellets are very promising for implementation in industrial production of xylose and glucose, xylitol, sorbitol, vegetable carbohydrate feeds, bioethanol, fodder yeast and food protein.

6 .The results of the research will be used in the next stage of work in the development of the process of physical and chemical characteristics, quantum-chemical studies and modeling .According to the results of the second stage of research there is a business partner SHF LLP NPF "Ana zher" to implement the technology of obtaining sorbitol and xylitol from cellulose guza-paya and beer pellet.

The overall objective of the project is technical substantiation and development of technology for deep processing of carbohydrate-containing raw materials to produce xylitol and sorbitol by means of chemical and enzymatic hydrolytic hydrolysis and hydrogenation of carbohydrate-containing plant raw materials and waste, based on the use of a combined (hybrid) process. It is known that the processing of hydrocarbon-containing raw materials consists of the following stages - pretreatment, hydrolysis and hydrogenation. According to the traditional technology it is necessary to perform numerous stages of purification, separation, intermediate processing, bringing the concentration of target products to the desired value. In the reporting year we have developed a single-reactor combined (hybrid) process of chemical and enzymatic hydrolysis and hydrogenation of cellulose of guza-paya and beer pellets. The optimum parameters of the process have been determined. The possibility of these processes in the presence of new generation catalysts developed by us, since traditional catalytic systems are inefficient in this direction, is shown. The results of our research and the opinions of scientists working in this direction, especially in the last decade, coincide in the fact that it is necessary to develop a flexible and not cumbersome technology for processing of hydrocarbon-containing raw materials and it is currently possible. Therefore, the idea of our project on combining the processes of hydrolysis and hydrogenation turned out to be very promising for the processing of hydrocarbon-containing raw materials, in particular, beer crush and guza-pai pulp. Application of this scheme allowed to achieve much higher yields at much lower costs, which once again proved the priority of this approach over multistage processes. We have a business partner to implement this development in production.

In order to fully understand the processes and catalysts developed by us, we further plan to investigate the physicochemical properties of contacts and identify the correlation between catalytic, adsorption and physicochemical properties, as well as to study the kinetics and mechanism of processes, quantum-chemical calculations and modeling. This will allow us to further understand the scientific and technical basis of combined (hybrid) processes.

REFERENCE LIST

1. Yang P., Kobayashi H., Fukuoka A. Recent Developments in the Catalytic Conversion of Cellulose into Valuable Chemicals // Chin. J. Catal. - 2011. - Vol. 32. - P. 716 - 722.

2. Terentyeva E.P., VΛOB6HKO H.K., Pavlova E.A., Aliev P.Γ. Fundamentals of chemistry of cellulose and wood: educational and methodical manual / GOUVPO SPbGTURP. - SP6, 2010. - 23 c.

3. Huber G.W., IborraS.,CoπτiaA. Synthesis of transportation fuels from biomass: Chemistry, catalysts, and engineering // Chem. Rev. - 2006. - Vol. 106. - P. 4044 - 4098.

4. A. Shrotri, A. Tanksale, J. N. Beltramini, H. Gurav andS. V. Chilukuri, Catal. Sci. Technol., 2012, 2, 1852-1858.

5. Y. Zhu, Z. N. Kong, L. P. Stubbs, H. Lin, S. Shen, E. V. Anslynand J. A. Maguire, ChemSusChem, 2010, 3, 67-70.

6. Palkovits R. Pentenoic acid pathways for cellulosic biofuels. // Angew. Chem. Int. Ed. - 2010. - Vol. 49. - No. 26.- P. 4336-4338.

7. Palkovits R., TajvidiK.,Procelewska J., Ruppert A. Efficient conversion of cellulose to sugar alcohols combining acid and hydrogenation catalysts. // From Abstracts of Papers, 241st ACS National Meeting & Exposition, Anaheim, CA, United States, March 27-3, 2011, CELL-240.

8. Palkovits R., TajvidiK.,Procelewska J., Rinaldi R. and Ruppert A. Hydrogenolysis of cellulose combining mineral acids and hydrogenation catalysts. // Green Chem. - 2010. - Vol. 12. - P. 972 - 978.

9. Fukuoka A. and Dhepe P. L. Catalytic Conversion of Cellulose into Sugar Alcohols // Angew. Chem. Ed. - 2007. - Vo1. 45. - P. 5161 -5163.

10. GeboersJ., S. Van de Vyver, Carpentier K., K. de Blochouse, Jacobs P. and Sels B.. Efficient catalytic conversion of concentrated cellulose feeds to hexitols with

heteropoly acids and Ru oπ carbon. // Chem. Commun. - 2010. - Vol. 46. - No. 20.- P. 3577 - 3579.

11. Geboers J., Van de Vyver S., Carpentier K., Jacobs P., Sels B.. Efficient hydrolytic hydrogenation of cellulose in the presence of Ru-loaded zeolites and trace amounts of mineral acid. // Chem. Commun. - 2011. - Vol. 47. - P. 5590 - 5592.

12. Method for obtaining glucose from cellulose-containing waste RU(11) 2 346 05513) C2

13. Method for obtaining xylitol from beer pellets (19) KZ (13) A4 (11) 30354

14. Palkovits R., TajvidiK.,Ruppert A. M. and Procelewska J. Heteropoly acids as efficient acid catalysts in the one-step conversion of cellulose to sugar alcohols. Chem. Commun. -201 l. - Vol. 47. - P.576-578.

15. Geboers J, Van de V. Stijn, Carpentier K., Blochouse K., Jacobs P., Sels B. Reductive splitting of concentrated cellulose feeds to hexitols with heteropoly acids and Ru op carbon.// From Preprints - American Chemical Society, Division of Petroleum Chemistry. - 201 1 . - Vol. 56. - No. 1.- P. 163.

16. Glinka N.L. General chemistry. Publisher: Integral-Press. ISBN 589602-017-1. - 2008. - C. 571.

17. Tian J., Wang J., Zhao S., Jiang C., Zhang X. and Wang X. Hydrolysis of cellulose by the heteropoly acid H PW, O40- °/ Cellulose. - 2010. - Vol. 17. - P.587-594.

18. Shimizu K., Furukawa H., Kobayashi N., Itaya U. and Satsuma A. Effects of Bronsted and Lewis acidities on the activity and selectivity of heteropolyacid-based catalyst for hydrolysis of cellobiose and cellulose. // Green Chem. - 2009. - Vol. 11. - P. 627-1632.

19. Barton D. G., Soled S. L. and Iglesia E. Acid Catalysts Based on Supported Tungsten Oxides. // Top. Catal. - 1998. - Vol. 6. - P. 87-99.

20. Kozhevnikov I.V. Catalysis by Heteropoly Acids and Multicomponent Polyoxometalates in Liquid-Phase Reactions. // Chem. Rev. - 1998. - Vol. 98. - P. 171-

198.

21. Catalyst for cellulose hydrolysis and/or recovery of cellulose hydrolysis products and method for producing sugar alcohols from cellulose

22. Catalyst for cellulose hydrolysis and/or recovery of cellulose hydrolysis products and method for producing sugar alcohols from cellulose

23. Method of saccharification of lignocellulosic raw material (RU 2405838)

24. S. Van de Vyver, GeboersJ., Jacobs P. and Sels B. Recent advances in the catalytic conversion of cellulose. // Chem.Cat.Chem. - 2011. - Vol. 3. - No. 1.-P. 82-94.

25. GeboersJ., S. Van de Vyver, Carpentier K., Jacobs P. and Sels B. Hydrolytic hydrogenation of cellulose with hydrotreatedcaesium salts of heteropoly acids and Ru/C. // Green Chem. - 2011. - Vol. 13. - P.2167- 2174.

26. Guo F., Fang Z., Xu C. S., Smith R. L. Solid acid mediated hydrolysis of biomass for producing biofuels. // Progress in Energy and Combustion Science. - 2012. - P. 1 - 19.

27. Lail D., Deng L., Lil J., Liao W., Guo Q., Fu W. Hydrolysis of Cellulose into Glucose by Magnetic Solid Acid. // CheinSusChem. - 2011. - Vol. 4. - No. 1. - P.55-58.

28. Toda M.,Takagaki A., Okamura M., Kondo J. N., Hayashi S., Domen K., Hara M.. Biodiesel made with sugar catalyst. // Nature. - 2005. - Vol. 438. P. 178.

29. Suganuma S., Nakajima K., Kitano M., Yamaguchi D., Kato H., Hayashi S., Hara M. Hydrolysis of Cellulose by Amorphous Carbon Bearing SOzH, COOH, and OH Groups. // J. Are. Chem. Soc. - 2008. - Vol. 130. - P. 12787.

30. Okamura M.,Takagaki A., Toda M., Kondo J. N., Domen K., Tatsumi T., Hara M., Hayashi S. Acid-catalyzed reactions of flexible polycyclic aromatic carbon in amorphous carbon. // Chem Mater. - 2006. - Vol. 18. - P. 30393045.

31. Yamaguchi D., Kitano M.,Suganuma S., Nakajima K., Kato H., Hara M. J. Hydrolysis of Cellulose by a Solid Acid Catalyst under Optimal Reaction ' Conditions.

// Phys. Chem. C. - 2009. - Vol. 113. - P. 3181-3188.

32. OndaA., Ochi T., Yanagisawa K. Selective hydrolysis of cellulose into glucose over solid acid catalysts. // Green Chem. - 2008. - Vol. 10. - P. 10331037.

33. C. Du, L. Zhou and S. Zhao, Adv. Mater. Res., 2011, 236-238,432-436.

34. S. Van de Vyver, J. Geboers, M. Dusselier, H. Schepers,T. Vosch, L. Zhang, G. Van Tendeloo, P. A. Jacobs andB. F. Sels, ChemSusChem, 2010, 3, 698-701.

35. OndaA., Ochi T.,Yanagisawa K. Hydrolysis of cellulose selectively into glucose over sulfonated activated-carbon catalyst under hydrothermal conditions. // Topic. Catalysis. - 2009. - Vol. 52. - P.801-807.

36. Strain of mycelial fungus aspergillusfoetidusbkm f 3890d - producer of acidic protease and carbohydrase complex containing pectinase (polygalacturonase), xylanase, p-glucanase, arabinase, galactanase, xyloglucanase, saccharase, a-l-arabinofuranosidase, p-glucosidase and amylase RU(11) 2 323 97313) C1

37. И. I. Fazliev, C. T. Minzanova, F. Y. Akhmadullina, P. 3. Musin, L. G. Mironova Prospects of beer pellets processing for xylose production.

38. Improved technology of bioethanol production was developed.

39. Dae Sung Park, Danim Yun, Tae Yong Kim, JayeonBaek, Yang Sik Yun, and Jongheop Yi. A Mesoporous Carbon-Supported Pt Nanocatalyst for the Conversion of Lignocellulose to Sugar Alcohols/ DOI: 10.1002/cssc.201300751/2013 Wiley-VCH Verlag GmbH & Co. KGaA, WeinheimChemSusChem 2013, 6, 2281 - 2289

40. Van de VyverS., Peng L., Geboers J., Schepers H., De Clippel F., Gommes C. J., Goderis B., Jacobs P. A., Sels B. F. Sulfonated silica/carbonnanocomposites as novel catalysts for hydrolysis of cellulose to glucose. //Green Chem. - 2010. - Vol. 12 - P. 1560-1563.

41. Pang J., Wang A., Zheng M., Zhang T. Hydrolysis of cellulose into glucose over carbons sulfonated at elevated temperatures. // ChemCommun. - 2010.- Vol. 46. - P. 6935-6937.

42. Joung Woo Han, Hyunjoo Lee, Joung Woo Han, Hyunjoo Lee .Direct conversion of cellulose into sorbitol using dual-functionalized catalysts inneutral aqueous solution. CatalysisCommunications 19 (2012) 255-260

43. Takagaakkii A.,T.TaaguussaagaawwaaC.,Domen K.. Glucose production fromsaccharides using layered transition metal oxide and exfoliated nanosheets as a water-tolerant solid acid catalyst. // ChemCommun. - 2008. - P. 5363- 5365.

44. Tagusagawa C., Takagaki A., Iguchi A., Takanabe K., Kondo J. N., Ebitani K., Tatsumi T., Domen K. Synthesis and Characterization of Mesoporous Ta-W Oxides as Strong Solid Acid Catalysts. // Chem Mater. - 2010. - Vol. 22. - P. 3072-3078.

45. Fukuoka A.,Dhepe P. L. Catalytic Conversion of Cellulose into Sugar Alcohols. // Angew. Chem. 2006. - Vol. 118. - P. 5285 - 5287.

46. Li R. X. Green Solvents: Synthesis and Application of lonic liquids. //China Chemical Industry Press, Beijing. - 2005. - P.298-300.

47. Zhu W., Widjaja E., Shirley L. P., Wang Z., Carpenter K., Maguire J. A., Hosmane N. S., Hawthome M. F. Ruthenium (0) nanoparticle-catalyzed isotope exchange between 10B and llB nuclei in decaborane (14). // J. Chem. Soc 2007. - Vol. 129. - P. 6507 - 6512.

48. Method of producing sugars from materials containing cellulose and hemicellulose, method of separating acid and sugars from liquids obtained by this method of production, method of digesting sugars obtained by this method of production, and method of processing solids obtained by this method of production (RU 2144087)

49. Method of ethanol production from cellulose-containing waste (variants) and method of glucose production from cellulose-containing waste (RU 2159816)

50. Tao F., Song H., Chou L. Catalytic conversion of cellulose to chemicals in ionic liquid. // Carbohydrate Research. - 2011. - Vol. 346, Issue l. P. 58-63.

51. Rinaldi R., Palkovits R., Schuth F. Depolymerization of cellulose by solid catalysts in ionic liquiAngew. Chem. - 2008. - Vol. 120. - P. 8167 - 8170.

52. DE 102008014. German patent.Depolymerization of cellulose by solid catalysts in ionic liquids / Rinaldi R., Palkovits R., Schuth F. - N DE10/ 2008/014/735.42008, international publication date 12.10.2008. - 16 P.

53. Rinaldi R., Maine J. N., J. von Stein, Palkovits R., Schfith F.What is the factor responsible for the control of depolymerization ofcellulose in ionic liquids - Solid acid catalyst or cellulose? // ChemSusChem. -2010. - Vol. 3. -P. 266276.

54. Swatloski R. P., Spear S. K., Holbrey J. D., Rogers R. D. Dissolution of Cellose with lonic Liquids. // J. Are. Chem. Soc. -2002. - Vo1. 124. -P. 4974 - 4975.

55. Zhang H., Wu J., Zhang J., ‰J.S. 1 -Allyl-3-methylimidazolium chloride hot temperature ionic liquid: A new and powerful nonderivatizing solvent for cellulose. // Macroimiolecules. 2005. - Vol. 38. - P.8272-8277.

56. WO 2012035160.International patent. Simultaneous hydrolysis and hydrogenation of cellulose / Li J. MakkeeM.,Moulijn J. A., O'connor P., Rasser J. C., Rosheuvel A. E. - N PCT/EP2O1 l/066156, priority date 17.09.2010; intemational publication date 22.03. 2012. - 20 P.

57. Zhu W., Kong Z. N., Stubbs L. P., Lin H., Shen S., Anslyn E. V., Maguire J. A. Conversion of Cellulose to Hexitols Catalyzed by Ionic Liquid- Stabilized Ruthenium Nanopanicles and a Reversible Binding Agent.// Chem. Sus. Chem. -2010. - Vo1. 3. - P. 67 - 70.

58. Kobayashi H., Matsuhashi H., Komanoya T., Haraa K., Fukuoka A. Transfer hydrogenation of cellulose to sugar alcohols over supported ruthenium catalysts. // Chem. Commun. -2011. - Vol. 47. -P. 23 66 - 2368.

59. Kuznetsov, B.N. Catalytic methods in obtaining chemical products from wood biomass // Chemistry for Sustainable Development.2007, Vol. 6, P. 383-396.

60. Kazeem Bode Olanrewaju Reaction kinetics of cellulose hydrolysis in subcritical and supercritical water. Iowa. 2012

61. Yakutseni B.П., Petrova Yu.E. Problems and practice of formation of state reserves of oil and gas // Oil and Gas Geology. Theory and practice. Electron.nauchn.

zhurnal.2009.No.4.

62. Azarov V.I., Burov A.V., Obolenskaya A.V. Microcrystalline cellulose. Chemistry of wood and synthetic polymers: textbook for universities. St. Petersburg, 1999. C.578-579.

63. Petropavlovskiy G. A., Kotelnikova H. E. Microcrystalline cellulose: a review / Chemistry of Wood.1979. №6 C. 3-21.

64. Schneider E.E. Hydrolysis production//ref. Sbornik. 1972. № 8. C.10 - 15

65. Corma A., Iborra S., Velty A. Chemical routes for the transformation of biomass into chemicals // Chem Rev.2007.Vol. 107.P. 2411 - 2502.

66. GalbraichL.S. Cellulose derivatives//Sorosovsky obrazovatel'nyjournal.1996.№11.C.47-53.

67. Fengel D., Wegener G. Wood (chemistry, ultrastructure, reactions). Moscow: Lesnaya Promyshlennaya Industriya, 1988.512 p.

68. JianrongLi, HelenaS. M. P. Soares, JacobA. MoulijnandMichielMakkee. Simultaneous hydrolysis and hydrogenation of cellobiose to sorbitol in molten salt hydrate media. Catalysis Science & Technology/ This journal is c The Royal Society of Chemistry 2013 Catal. Sci. Technol., 2013, 3, 1565-1572

69. Ning Yan, Chen Zhao, Chen Luo, Paul J. Dyson, Haichao Liu, and Yuan Kou One-Step Conversion of Cellobiose to C6-Alcohols Using a RutheniumNanocluster Catalyst. C O M M U N I C A T I O N S J. AM. CHEM. SOC. 9 VOL. 128, NO. 27, 2006 8715 Published on Web 06/20/2006

70. Joung Woo Han, Hyunjoo Lee, Joung Woo Han, Hyunjoo Lee .Direct conversion of cellulose into sorbitol using dual-functionalized catalysts inneutral aqueous solution. Catalysis Communications 19 (2012) 115-118

71. Jianrong Li, Helena S. M. P. Soares, Jacob A. Moulijn and MichielMakkee. Simultaneous hydrolysis and hydrogenation of cellobiose to sorbitol in molten salt hydrate media. Catalysis Science & Technology/ This journal is c The Royal Society

of Chemistry 2013 Catal. Sci. Technol., 2013, 3, 1565-1572

72. H.H. Trofimova, V.A. Babkin Study of acid hydrolysis of larch wood polysaccharides for obtaining crystalline glucose Chemistry of plant raw materials. 2009. №3. C. 31-37.

73. Outlov S. A., Bazarnova HT., Kushnir E. Yu. Microcrystalline cellulase: structure, properties and applications (review) // Chemistry of plant raw materials. 2013. №3.C.33-41.

74. 4 .ArdizzoneS., DioguardiF.S., MussiniT. ,MussiniP.R.,RondininiS.,VercelliB .,VertovaA.Microcrystallinecellulosepowders:structure,surface features and water sorption capability // Cellulose. 1999. Vol, 6. N1.P. 57-69.

75. Advances in Chemistry and Chemical Technology. Vol. XXIX. 2015. № 7O.V. Manaenkov, O.V. Kislitsa, A.E. Filatova/ Catalysts based on sps for hydrolytic hydrogenation of cellulose

76. Ning Yan, Chen Zhao, Chen Luo, Paul J. Dyson, Haichao Liu, and Yuan Kou One-Step Conversion of Cellobiose to C6-Alcohols Using a RutheniumNanocluster Catalyst. C O M M U N I C A T I O N S J. AM. CHEM. SOC. 9 VOL. 128, NO. 27, 2006 8715 Published on Web 06/20/2006

77. O.V. Manaenkov, O.V. Kislitsa, A.E. Filatova / Sps-based catalysts for hydrolytic hydrogenation of cellulose

78. Torpolov M.A., Tarabukin D.V., Frolova S.V., Shcherbakova T.P., Volodin V.V. Enzymatic hydrolysis of powdered celluloses obtained by different methods. // Chemistry of plant raw materials. 2007. №3. C. 69-76.

79. Budayeva V.V., Mitrofanov R.Yu., Zolotukhin V.N., Obrezkova M.V., Skiba E.A., Ilyasov S.G., Sakovich G.V., Oparina L.A., Vysotskaya O.V., Kolyvanov N.A., Gusarova N.K., Trofimov B.A. Ways of complete and environmentally friendly processing of renewable plant raw materials // Polzunov Bulletin.2010.№ 4-1.P. 158 - 167.

80. Method of catalytic conversion of cellulose into hexitols (RU 2497800)

81. Vratislav T. Glucose hydrogenation in a trick-bed reactor // Collection of Czechoslovak Chemical Communication. 1997. V. 62. pp. 1423 - 1428.

82. Sabine S., Catherine L., Peter C. Ni/Si02 catalysts prepared with ethylenedi amine nickel precursors: Influence of the pretreatment on the catalytic properties in glucose hydrogenation // Applied Catalysis, A: General. 2007. V. 318. pp. 45-53.

83. Subramani V. and Gangwal S. K. Hydrogenation of Glucose to Sorbitol over nanoparticle Ni7A^Os catalyst // North American Catalysis Society (NACS) - 20th North American Meeting (NAM). 2007.

84. Velu S., Gangval S.K. and Nelson T.O. Highly dispersed nickel hydrogenation catalysts and method for making the same. 2006. Patent US No. 60849479.

85. Kusserow B., Schimpf S., Glaus P. Hydrogenation of Glucose to Sorbitol over Nickel and Ruthenium Catalysts // Adv. Synth. Catal. 2003. V. 345. pp. 289299.

86. Valange S., Barrault J., Derouault A., Gabelica Z. Binary Cu-Al mesophases precursors to uniformly sized copper particles highly dispersed on mesoporous alumina // Micropor. Mesopor.Mater. 2001. V. 44 - 45. pp. 211 -220.

87. Joung Woo Han, Hyunjoo Lee, Joung Woo Han, Hyunjoo Lee .Direct conversion of cellulose into sorbitol using dual-functionalized catalysts inneutral aqueous solution. Catalysis Communications 19 (2012) 115-118

88. Qiying Liu, Jin Tan, ChiliuCai, Longlong Ma, and TiejunWang .Enhanced Sugar Alcohol Production from Cellulose by Pretreatment with Mixed Ball-Milling and Solid Acids. BIORESOURCES.COM.

89. Dechhamp N., Gamez A., Pezzard A., Gallezot P. Kinetics of glucose hydrogenation in a trick-bed reactor // Catalysis Today. 1995. V. 24. pp. 29 - 34.

90. Guo A., Li H., Zhu J., Ye W., Qiao M., Dai W. Liquid phase glucose hydrogenation to d-glucitol over an ultrafme Ru-B amorphous alloy catalyst // Journal of Molecular Catalysis A: Chemical. 2003. V. 200. pp. 213-221.

91. Li H., Li H., Jing-Fa D. Glucosehydrogenation over Ni - B/Si02 amoφhous alloy

catalyst and the promoting effect of metal dopands // Catalysis Today. 2002. V. 74. pp. 53-63.

92. Li H., Wang W., Deng J.-F. Glucose Hydrogenation to Sorbitol over a Skeletal Ni-P Amorphous Alloy Catalyst (Raney Ni-P) // J. Geophys. Catal. 2000. V. 191. pp. 257 - 260.

93. Gallezot P., Cerino P.J., Blanc B., Fleche G., Fuerters P. Glucose hydrogenation on promoted raney-nickel catalysts // J. Catal. Catal. 1994. V. 146. pp. 93 - 102.

94. Wisniak J., Simon R. Hydrogenation of Glucose, Fructose, and Their Mixtures // Ind. Eng. Chem. Prod. Res. Dew. 1979. V. 18. pp. 50 - 57.

95. Perrard A., Gallezot P., Joly J.-P. Higly efficient metal catalysts suppor on activated carbon cloths: A catalytic application for the hydrogenation of D-glucose to D-sorbitol // App. Cat. A: General. 2007. V. 331. pp. 100- 104.

96. Method of obtaining xylose and cellulose for chemical processing

97. Method of hydrolysis of plant fibrous material to produce and isolate a saccharide including glucose

98. Owners of patent RU 2461633

99. Method for obtaining xylitol from aqueous xylose solution

100. H. Kobayashi, Y. Ito, T. Komanoya, Y. Hosaka, P. L. Dhepe,K. Kasai, K. Hara and A. Fukuoka, Green Chem., 2011, 13,326-333.

101. R. Palkovits, K. Tajvidi, A. M. Ruppert and J. Procelewska,Chem. Commun.,2011, 47, 576-578.

102. B. W. Hoffer, E. Crezee, P. R. M. Mooijman, A. D. vanLangeveld, F. Kapteijn and J. A. Moulijn, Catal. Today,2003, 79-80, 35-41.

103. P. Gallezot, N. Nicolaus, G. Fleche, P. Fuertes, andPerrard, J. Catal. 1998, 180, 51-55.

104. H. Wang, L. Zhu, S. Peng, F. Peng, H. Yu and J. Yang,Renewable Energy, 2012, 37, 192-196.

105. W. Deng, X. Tan, W. Fang, Q. Zhang and Y. Wang, Catal.Lett., 2009, 133, 167-174.

106. J. Geboers, S. Van de Vyver, K. Carpentier, P. Jacobs andB. Sels, Chem. Commun., 2011, 47, 5590-5592.

More Books!

yes

I want morebooks!

Buy your books fast and straightforward online - at one of world's fastest growing online book stores! Environmentally sound due to Print-on-Demand technologies.

Buy your books online at

www.morebooks.shop

Kaufen Sie Ihre Bücher schnell und unkompliziert online – auf einer der am schnellsten wachsenden Buchhandelsplattformen weltweit! Dank Print-On-Demand umwelt- und ressourcenschonend produzi ert.

Bücher schneller online kaufen

www.morebooks.shop

info@omniscriptum.com
www.omniscriptum.com

OMNIScriptum

MIX
Papier aus verantwortungsvollen Quellen
Paper from responsible sources
FSC® C105338

Printed by Books on Demand GmbH, Norderstedt / Germany